Industrial Microbiology and
the Advent of Genetic Engineering

A **SCIENTIFIC** *Book*
AMERICAN

Industrial Microbiology and
the Advent of Genetic Engineering

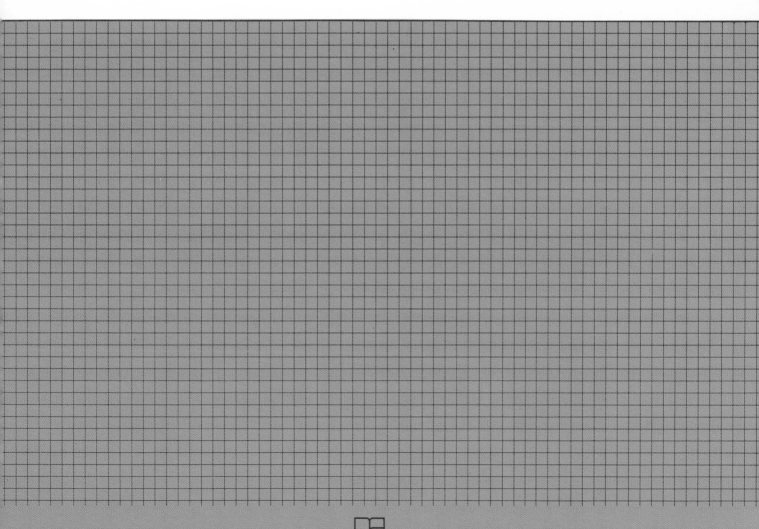

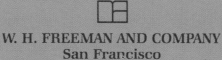

W. H. FREEMAN AND COMPANY
San Francisco

The Cover

The painting on the cover symbolizes the theme of this book: industrial microbiology and the advent of genetic engineering. The tank at the left is a fermenter, the type of vessel in which the microorganisms that manufacture the products of industrial microbiology are grown. The piping associated with the vessel provides services to satisfy the requirements of the microorganism. The red pipes carry steam to sterilize the assembly and the growth medium before an inoculum of the organism is introduced into the vessel. The green pipes carry cold or steam-heated water to control the temperature of the culture. The black pipes carry the air needed by the organism. (Some of the organisms of industrial microbiology are anaerobic: they grow in the absence of air and do not need any.) The medium in the fermenter is stirred by a motor-driven agitator. The fermenter in the painting is made by the New Brunswick Scientific Co. Its capacity is 250 liters, so that it is suitable for pilot-plant work. Fermenters with a capacity of 50,000 gallons (about 200,000 liters) are common in industry.

Cover painting by Ted Lodigensky.

Library of Congress Cataloging in Publication Data

Main entry under title:

Industrial microbiology and the advent of genetic engineering.

"A Scientific American book."
"The eight chapters . . . originally appeared as articles in the September 1981 issue of Scientific American."—T.p. verso.
Bibliography: p.
Includes index.
1. Industrial microbiology—Addresses, essays, lectures.
I. Scientific American.
QR53.I525   660'.62   81-15244
ISBN 0-7167-1385-3   AACR2
ISBN 0-7167-1386-1 (pbk.)

The eight chapters in this book originally appeared as articles in the September 1981 issue of *Scientific American*.

Printed in the United States of America

1234567890 DO 8987654321

Contents

Foreword

Bacteria, yeasts and molds are machines ideally equipped to do manufacturing on a molecular scale. Indeed, they are self-replicating machines: once set in motion, they not only make the desired product but also assemble copies of themselves. Industrial microbiology is the enterprise that exploits such biological machines to work the will of men. It is an established technology now being transformed by the new methods of genetic engineering that bring the manufacturing capabilities of the microorganism more directly under human control.

Up to now the only organisms available for industrial purposes have been those obtained by selection from natural populations. Over many generations, selection can have a powerful effect, and certain organisms have been induced to divert a major share of their metabolic energy into the production of substances more useful to man than to the organism in its natural setting. Nevertheless, the inescapable limitation of selection is that it can act only on organisms that already exist, favoring those whose genome, or total complement of genes, specifies the synthesis of desired materials. Selection can never introduce new information into the genome.

The new methods of genetic engineering give the biologist direct access to the genome. A gene can be inserted, deleted, altered or duplicated. For example, a gene from one organism can be transferred into the genome of another, and such a transfer of genetic information can be accomplished even across the evolutionary distance that separates a human being from a bacterium. An artificial gene, encoding instructions for making a molecule unknown in nature, can be assembled by chemical means and introduced into a microorganism. Two factors have made possible this editing of the genome. One factor is the remarkable universality of the genetic code: the message embodied in a length of DNA is interpreted in exactly the same way by virtually all living organisms. The second factor is the ingenuity of biologists in deciphering the code and in reconstructing the mechanisms by which the cell reads it and acts on it.

Before the potential of genetic engineering can become a reality, these techniques must be taken out of the glassware of the laboratory and put to work in the steel vessels of industry. What has not been fully appreciated by all observers of the recent developments in molecular biology is that an industrial context for genetic engineering already exists; microorganisms have long made an important contribution to the world economy. The origins of the technology are in the ancient arts of making bread and beer. In the early years of this century, methods borrowed from brewing and baking were applied to the production of industrial chemicals such as solvents and alcohols. In the past forty years these methods have been adapted to the synthesis of antibiotics and other pharmaceuticals. The scale of such undertakings is large. Batches of 50,000 liters are now commonplace, and new microbiological technology may soon allow the processing of raw materials into products in a continuous stream, as is done through chemical technology in a petroleum refinery.

The task of incorporating the techniques of genetic engineering into industrial practice is already under way. For example, plans have been announced for the commercial production of human insulin in the bacterium *Escherichia coli*, an organism that is native to the human colon but now is very much at home in the laboratory and soon will have a place in industry.

In the chapters that follow, the accomplishments and the promise of genetic engineering and industrial microbiology are set forth by its practitioners. All of this material first appeared in the September 1981 issue of SCIENTIFIC AMERICAN, which was the thirty-second in the series of single-topic issues published annually by the magazine. It has been made available in book form through the expeditious efforts of W. H. Freeman and Company, the book-publishing affiliate of SCIENTIFIC AMERICAN.

THE EDITORS*

September 1981

*BOARD OF EDITORS: Gerard Piel (Publisher), Dennis Flanagan (Editor), Brian P. Hayes (Associate Editor), Philip Morrison (Book Editor), Francis Bello, John M. Benditt, Peter G. Brown, Michael Feirtag, Paul W. Hoffman, Jonathan B. Piel, John Purcell, James T. Rogers, Armand Schwab, Jr., Joseph Wisnovsky.

1

Industrial Microbiology

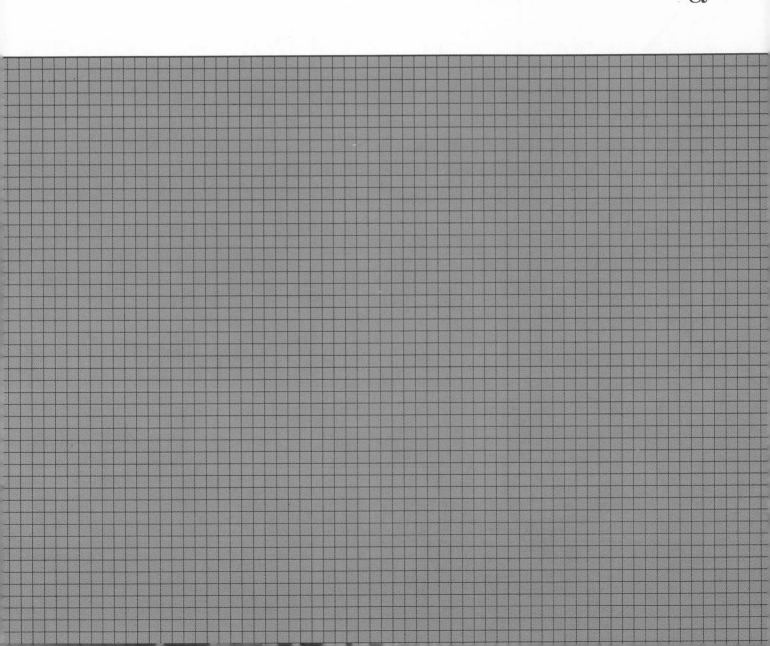

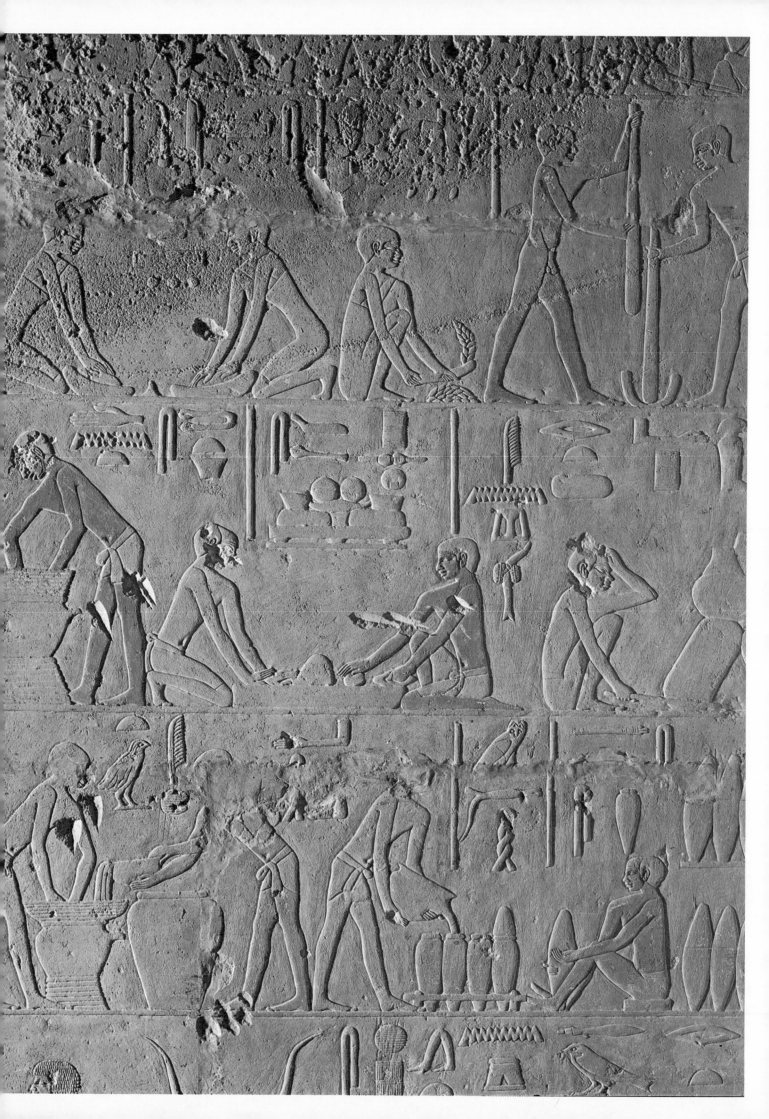

Industrial Microbiology

BY ARNOLD L. DEMAIN AND NADINE A. SOLOMON

Introducing a volume on the making of food, drink, pharmaceuticals and industrial chemicals by microorganisms, with special reference to newer methods of programming the microorganisms for their task

Industrial microbiology is in a ferment, a condition nontechnically defined as one of agitation, turbulence or general unrest. Recent advances in molecular biology have generated a wave of excitement about the prospective application of novel microbiological techniques in a wide range of industrial roles. What is often lacking in public discussions of recombinant DNA, genetic engineering and the like, however, is a sense of the context in which the new developments are to take place. Industrial microbiology is not just a new field of entrepreneurial activity; it is a well-established factor in the world economy, responsible for a current annual production valued at tens of billions of dollars in the U.S. alone. Moreover, it is the outgrowth of a pervasive human activity with a rich history that goes back thousands of years.

This *Scientific American* book is devoted to industrial microbiology, with special reference to the changes in it that are likely to result from the introduction of the new tools of genetic manipulation. The articles that follow will address major subdivisions of the topic, in each case placing the anticipated bene-

fits of the emerging biotechnology in the appropriate historical, economic and social perspective. First, however, we shall present a brief overview of the entire field.

The art of fermentation, technically defined in its broadest sense as the chemical transformation of organic compounds with the aid of enzymes (particularly those made by microorganisms), is very old. The ability of yeast to make alcohol in the form of beer was known to the Sumerians and the Babylonians before 6000 B.C. Much later, by about 4000 B.C., the Egyptians discovered that the carbon dioxide generated by the action of brewer's yeast could leaven bread. Reference to wine, another ancient product of fermentation, can be found in the Book of Genesis, where it is noted that Noah consumed a bit too much of the beverage.

By the 14th century A.D. the distillation of alcoholic spirits from fermented grain, a practice thought to have originated in China or the Middle East, was common in many parts of the world. Other fermentation processes with their roots deep in antiquity include the culti-

vation of acetic acid bacteria to make vinegar, lactic acid bacteria to preserve milk (for example in the form of yogurt) and various bacteria and molds to produce cheese.

Microorganisms provided food and drink for more than 8,000 years before their existence was recognized in the 17th century. Then the pioneering Dutch microscopist Anton van Leeuwenhoek, turning his simple lens to the examination of water, decaying matter and scrapings from his teeth, reported the presence of tiny "animalcules," moving organisms less than a thousandth the size of a grain of sand. Many people thought such organisms arose spontaneously from nonliving matter. Although the theory of spontaneous generation, which had been postulated by Aristotle among others, was by then discredited in its application to higher forms of life, it did seem to explain how a clear broth became clouded by large numbers of such organisms as the broth aged. It was not until the second half of the 19th century that Louis Pasteur of France and John Tyndall of Britain demolished the concept of spontaneous generation and proved that existing microbial life comes from preexisting life.

Even before Pasteur began his work on the origin of microbial life three independent investigators, Charles Cagniard de la Tour of France and Theodor Schwann and Friedrich Traugott Kützing of Germany, had proposed that the products of fermentation, chiefly ethanol (ethyl alcohol) and carbon dioxide, were created by a microscopic form of life. This concept was bitterly opposed by the leading chemists of the period (men such as Jöns Jakob Berzelius, Justus von Liebig and Friedrich Wöhler), who believed fermentation was strictly a chemical reaction; they maintained that the yeast in the fermenta-

ANTIQUITY OF FERMENTATION is represented graphically by the scenes of baking and brewing depicted in the painted relief on the opposite page. The relief appears on the wall of a Fifth Dynasty Egyptian tomb dating from about 2400 B.C. It is now in the collection of the National Museum of Antiquities in Leiden. The figures in the top panel are engaged (*right to left*) in pounding, winnowing and grinding the grain (presumably barley or emmer, a primitive variety of wheat). Those in the middle panel are soaking the coarse-ground flour in water, allowing some of the whole grains to malt, or sprout (*left*), kneading the leavened dough and fashioning it into loaves of various shapes (*center*) and baking the "beer bread" in an oven (*right*). The baker is portrayed in a characteristic attitude, raking the fire with a long-handled implement held in one hand and shielding his eyes from the heat with the other. In the bottom panel the mash is shown being strained into a fermenting vat, which rests on a stand resembling a coiled rope. After fermenting for a few days the finished beer is poured from the vat into pottery jars, which are promptly capped, sealed with clay and placed in storage. The Egyptian brewers relied initially on yeasts in the air or on the skin or husk of fruits and cereals; later a pure or almost pure yeast became available. The ancient breweries produced several types of beer; some brands, said to be "strong," may have had an alcohol content as high as 12 percent.

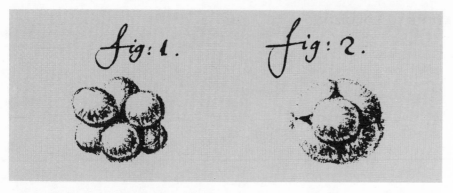

EARLIEST-KNOWN DRAWING of the apparent structure of a clump of yeast cells was enclosed in a letter sent by Anton van Leeuwenhoek in 1680 to Thomas Gale, at that time secretary of the Royal Society of London. On examining a sample of cold beer through his primitive microscope, van Leeuwenhoek noted a great many small particles. "Some of these," he later wrote (in Latin), "seemed to me to be quite round, others were irregular, and some exceeded the others in size and seemed to consist of two, three or four of the aforesaid particles joined together. Others again consisted of six globules and these last formed a complete globule of yeast.... With a view to representing this combination to the eye I took six globules of wax and joined them together as in Figure 1, and arranged them and had them drawn in such a way that all six could be seen at once. I next squeezed these aggregated globules in my hands so that they assumed the form shown in Figure 2, for I imagine that the result which I effected by rolling the globules of wax between my hands to compress them was nearly the same as in the fermentation of beer." Although van Leeuwenhoek was never able to actually distinguish the six cells that made up the larger globule, he reported that his observations, interpreted in the light of his wax models, "were as clear to me as if I had before my naked eye a very small transparent bubble that was filled with six other smaller ones." What he saw (at a magnification estimated to be about 125 diameters) was probably an aggregate of yeast cells formed by rapid budding.

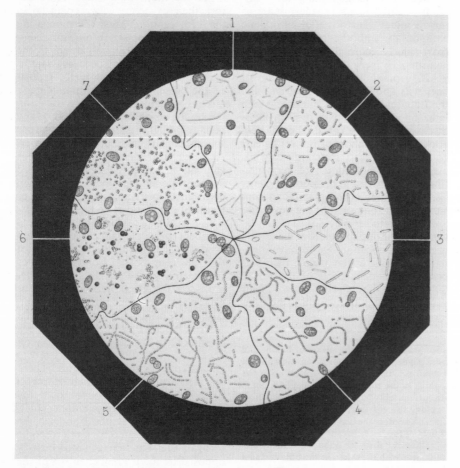

VARIOUS MICROORGANISMS responsible for the spoilage of beer were investigated by Louis Pasteur in his classic 19th-century study of brewing. This composite drawing, showing the principal microbial contaminants of beer and malt wort, is reproduced from the original 1876 French edition of his book *Études sur la bière, ses maladies, causes qui les provoquent. Procédés pour la rendre inaltérable, avec une théorie nouvelle de la fermentation* (*Studies of beer, its diseases and the causes that provoke them. Procedures for making it unalterable, with a new theory of fermentation*). It was in the course of this project that Pasteur demonstrated the existence of anaerobic, or airless, microbial life. Larger, more globular particles are yeasts.

tion broth was lifeless, decaying matter. Organic chemistry was flourishing at the time, and the opponents of the microbial hypothesis were initially quite successful in putting forth their views.

It took almost two decades—from 1857 to 1876—for Pasteur to disprove the chemical hypothesis. He had been called on by the distillers of Lille to find out why the contents of their fermentation vats were turning sour. He noted through his microscope that the fermentation broth contained not only yeast cells but also bacteria that could produce lactic acid. His greatest contribution during this 20-year period was to establish that each type of fermentation is mediated by a specific microorganism. Furthermore, in a study undertaken to determine why French beer was inferior to German beer, he demonstrated the existence of strictly anaerobic life: life in the absence of air.

One of Pasteur's central concepts, that each fermentation provides energy to the species conducting it, led to the accidental discovery of cell-free metabolism by Eduard Buchner of Germany in 1897. Buchner found that an extract of macerated yeast, freed of intact cells by filtration, retained the ability to convert sugar into alcohol. His discovery gave rise to the field of biochemistry. Later work showed that the biological conversion actually consisted of a series of simple chemical reactions, each catalyzed by a specific enzyme.

Considerable progress was made in basic biochemical research following Buchner's lead, but little of the new knowledge was applied to the practice of industrial fermentation until World War I. On the German side glycerol for the manufacture of explosives soon became an urgent need. The British naval blockade had interfered with the importation of vegetable oils, the usual raw material for the production of glycerol. Several years earlier Carl Neuberg, a German biochemist, had followed up on an observation of Pasteur's that small amounts of glycerol were usually produced in alcoholic fermentation. Neuberg discovered that the addition of sodium bisulfite to the fermentation vat favored the production of glycerol at the expense of ethanol. Although this prewar finding was thought to be of academic interest only, the Germans quickly developed it into an industrial fermentation yielding 1,000 tons of glycerol per month.

On the British side acetone was needed for the manufacture of munitions. In response to the shortage of the chemical Chaim Weizmann, a Russian-born chemist, developed the acetone-butanol fermentation, which depends on the anaerobic bacterium *Clostridium acetobutylicum*. (Weizmann was later the first president of Israel.) The German glycerol process passed out of the picture at the end of World War I, but the acetone-butanol process remained an important source of acetone for many years, until it was displaced by processes based on petroleum. It was the first large-scale fermentation for which problems of contamination by other bacteria and by bacteriophages (viruses that infect bacteria) had to be solved. For the first time pure-culture methods had to be employed in industrial fermenters, an experience that proved to be invaluable when the antibiotic era arrived in the 1940's.

For thousands of years moldy cheese, meat and bread had been employed in folk medicine to heal wounds. It was not until the 1870's, however, that Tyndall, Pasteur and William Roberts, a British physician, directly observed the antagonistic effects of one microorganism on another. Pasteur, with his characteristic foresight, suggested that the phenomenon might have some therapeutic potential. For the next 50 years various microbial preparations were tried as medicines, but they were either too toxic or inactive in live animals. Finally in 1928 Alexander Fleming noted that the mold *Penicillium notatum* killed his cultures of the bacterium *Staphylococcus aureus* when the mold accidentally contaminated the culture dishes. After growing the mold in a liquid medium and separating the fluid from the cells he found that the cell-free liquid could inhibit many species of bacteria. He gave the active ingredient in the liquid the name penicillin but soon afterward discontinued his work on the substance.

Attempts to isolate penicillin were made in the 1930's by a number of British chemists, but the instability of the substance frustrated their efforts. Eventually a study begun in 1939 at the University of Oxford by Howard W. Florey, Ernst B. Chain and their colleagues led to the successful preparation of a stable form of penicillin and to the demonstration of its remarkable antibacterial activity, first in experimental animals and then in man. Florey and his colleague Norman Heatley realized that conditions in wartime Britain were not conducive to the development of an industrial process for producing the antibiotic. They therefore came to the U.S. in 1941 to seek assistance. With the help of the U.S. Department of Agriculture and several American pharmaceutical companies the production of penicillin by a related mold, *Penicillium chrysogenum,* soon became a reality.

The advent of penicillin, which signaled the beginning of the antibiotic era, was closely followed by the discoveries of Selman A. Waksman, a soil microbiologist at Rutgers University, who succeeded in obtaining a number of new antibiotics from the class of microorganisms called actinomycetes; the best-known of his new "wonder drugs" was streptomycin. From Waksman's time up

PORTRAIT OF PASTEUR was made in 1884 on the occasion of a visit to Copenhagen. Pasteur was 61 at the time. He died in 1895. The photograph, which was made by J. Petersen & Son, is now in the archives of the Pasteur Museum, a part of the Pasteur Institute in Paris.

to the present there has been a proliferation of economically viable fermentation processes and products.

Underlying the diversity of microbial processes and products described in the following articles are certain characteristics shared by all microorganisms. The most fundamental is the small size of the microbial cell and its correspondingly high surface-to-volume ratio, which facilitates the rapid transport of nutrients into the cell and thereby supports its high metabolic rate. For example, the rate of production of protein in yeast is several orders of magnitude higher than it is in the soybean plant, which in turn is 10 times higher than it is in cattle. The extremely high rate of microbial biosynthesis enables some microorganisms to reproduce in only 15 minutes.

The environments that support microbial life reflect the broad spectrum of microbial evolution. Microorganisms have been found living at temperatures ranging from the freezing point of water to almost the boiling point, in salt water and fresh water, and in the presence of air and the absence of air. Some have evolved life cycles that include a stage of suspended animation in response to the depletion of nutrients: in the form of spores they may remain inactive for years until the environment becomes more favorable for growing cells.

Because microorganisms are capable of a wide variety of metabolic reactions they can adapt to many sources of nutrition. This adaptability makes it possible for industrial fermentations to rely on inexpensive nutrients. For example, molasses and cornsteep liquor, waste products respectively of the crystallization of sugar and the wet milling of corn, are both valuable for the production of penicillin.

There are four classes of industrially important microorganisms: yeasts, molds, single-cell bacteria and actinomycetes. The yeasts and the molds are more highly developed; together they constitute the fungi. Organisms of this type are eukaryotic, that is, their cells, like the cells of plants and animals, have a membrane-enclosed nucleus and more than one chromosome; they also contain organelles such as mitochondria (the tiny sausage-shaped bodies that are responsible for the cell's main energy supply). The single-cell bacteria and the actinomycetes, in contrast, are prokaryotic: they have no nuclear membrane or mitochondria, and they have only one chromosome. In addition the cells of prokaryotes are typically much smaller than those of eukaryotes. In spite of these basic biological differences there is a superficial resemblance between the molds and the actinomycetes in that both are filamentous: they grow as a branched system of threadlike hyphae rather than as single cells. The yeasts and the bacteria, on the other hand, are unicellular under normal conditions.

The commercially important prod- ucts of these microorganisms fall into four major categories: (1) the microbial cells themselves; (2) the large molecules, such as enzymes, that they synthesize; (3) their primary metabolic products (compounds essential to their growth), and (4) their secondary metabolic products (compounds not required for their growth). In general both the primary and the secondary metabolites of commercial interest have a fairly low molecular weight: less than 1,500 daltons, compared with the molecular weight

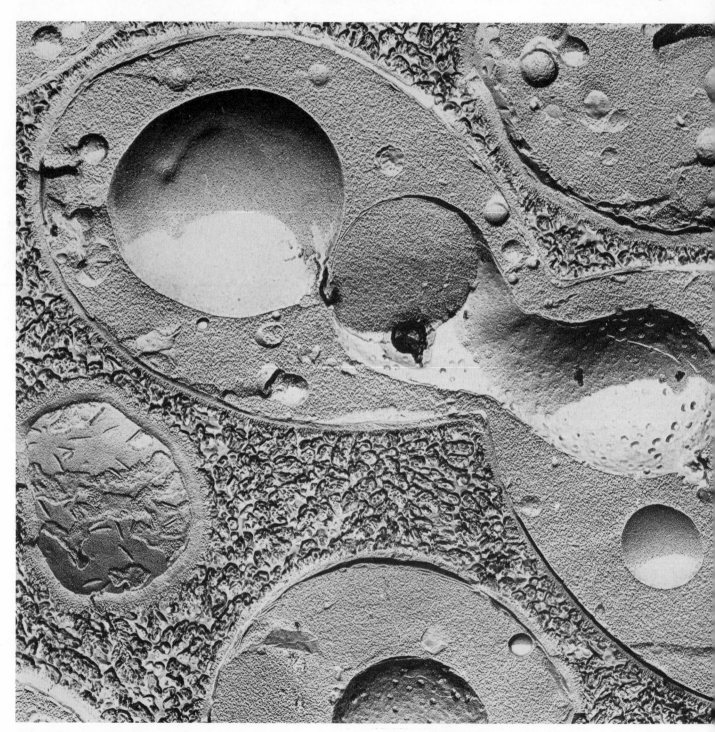

DIVIDING YEAST CELLS appear in an electron micrograph made by the freeze-fracture-etching technique. The cells belong to the species *Saccharomyces cerevisiae* (brewer's yeast). The specimen was first frozen and then fractured; the ice matrix was next etched away slightly, leaving the material to be replicated in relief. (The electron micrograph is actually made from a thin-film replica of the surface.) The fracture split the cells open, revealing their interior structure. The smaller dumbbell-shaped area inside the larger dumbbell-shaped one is the dividing nucleus of one of the cells. Pores can be seen on the surface of the nuclear membrane. The other shapes visible

of an enzyme, which can range from 10,000 to several million daltons.

Microbial cells have two main commercial applications. The first is as a source of protein, primarily for animal feed. In its commonest form this product is referred to as single-cell protein, although in fact it usually includes the entire microbial cell, the major component of which is protein.

Microbial cells are also used to carry out biological conversions, processes in which a compound is changed into a structurally related compound by means of one or more enzymes supplied by the cells. Biological conversions, also known as microbial transformations, can be accomplished with growing cells, nongrowing cells, spores or even dried cells. Microorganisms, which can carry out almost every kind of chemical reaction, have many advantages over chemical reagents. For example, many nonbiological chemical reactions call for a considerable input of energy to heat or cool the reaction vessel; in addition they are generally conducted in solvents and require inorganic catalysts, both of which may be pollutants. Finally, many nonbiological chemical reactions yield unwanted by-products that must be removed in a separate purification step.

Unlike most nonbiological chemical reactions, biological conversions proceed at biological temperatures with water as the solvent. The cells can often be immobilized on a supporting structure for continuous processing. Another valuable asset of biological conversions is their specificity: one enzyme usually catalyzes only one kind of reaction at a specific site on the substrate molecule. The enzyme can also be made to select one isomer, or molecular form of a compound, in a mixture of forms to produce a single isomer of the product. These characteristics account for the high yields typical of biological conversions, which can reach 100 percent.

The biological conversion of ethanol into a dilute solution of acetic acid (vinegar) was done in Babylon by 5000 B.C. Other important biological conversions transform isopropanol into acetone, glucose into gluconic acid and sorbitol into sorbose. (The last reaction is the only biological step in the otherwise nonbiological manufacture of ascorbic acid, vitamin C.) Among the notable biological conversions in the pharmaceutical industry are those involved in the production of steroids. More recently, in the production of semisynthetic penicillins it became possible to replace a chemical reaction that creates pollutants with a nonpolluting biological conversion.

The most versatile large molecules manufactured by microorganisms are enzymes. These biological catalysts are important in the food and chemical industries because of their specificity, efficiency and potency under conditions of moderate temperature and acidity. Although enzymes have traditionally been extracted from plants and animals, their production by microorganisms is increasing rapidly owing to the increasing availability of such organisms and the ease with which the yield can be improved by manipulating either the genes or the environment of the organisms. Moreover, in the microbial production of enzymes the fermentation times are short, the growth mediums are inexpensive and the screening procedures are simple.

Recent applications of microbially produced enzymes include the use of amylases in brewing, baking and the manufacture of textiles; of proteases in brewing, meat tenderizing and the manufacture of detergents and leather, and of rennin in cheesemaking. A major recent development has been the combination of three microbially produced enzymes—alpha-amylase, glucamylase and glucose isomerase—to obtain a high-fructose sweetening agent from cornstarch. At present there is great interest in enzymes immobilized on a solid substrate, which offer many advantages over free enzymes.

In the biosynthesis of enzymes it is often necessary to exploit or bypass certain regulatory mechanisms that have evolved over millions of years to prevent the overproduction of enzymes and their products. For example, the production of an enzyme can be enhanced by a factor of 1,000 by adding a special inducer substance to the fermentation vat; the inducer may be either the substrate on which the enzyme acts or a compound structurally similar to the substrate. The manufacture of an enzyme is sometimes repressed by a natural feedback mechanism associated with the end product of the metabolic pathway in which the enzyme functions; the repression can be avoided by limiting the accumulation of that particular end product in the cell. Another important type of repression, called catabolite repression, is avoided by replacing rapidly utilized sources of carbon and nitrogen (such as glucose and ammonia) with nutrients such as starch or soybean meal, which are consumed more slowly.

Besides enzymes the class of commercially important large molecules that can be made by microorganisms includes polysaccharides (long-chain molecules consisting of repeating sugar units). For years the major source of polysaccharides for industry has been plants, particularly seaweeds. Recently, however, there has been renewed interest in polysaccharides manufactured by microorganisms. Of the thousands of different polysaccharides that can be microbially produced the best-known is xanthan, which is made by the bacterium *Xanthomonas campestris*. This colloidal substance is added to many foods as a stabilizer and thickener, and the petroleum industry is beginning to include it as an ingredient in drilling muds. Among the other important large molecules that can be obtained from microorganisms are the active ingredients of vaccines and insecticides.

In order to produce primary metabolites commercially by fermentation the regulatory mechanisms that govern the synthesis of enzymes and their activity in microorganisms must be bypassed.

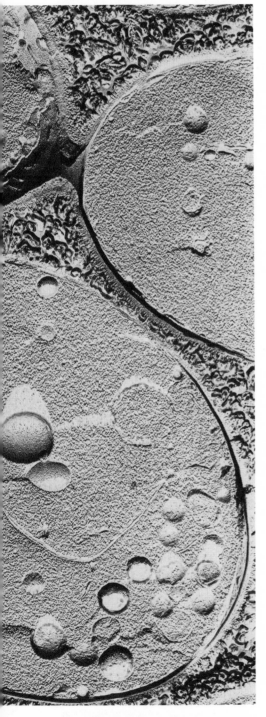

in the cells are either vacuoles (spherical cavities) or mitochondria (cellular organelles). The micrograph, which enlarges cells some 21,000 diameters, was made by H. Moor of the Swiss Federal Institute of Technology.

These mechanisms evolved to regulate enzymatic reactions because it is usually detrimental for an organism in nature to overproduce its internal metabolites. If it does, it merely secretes them into the environment for other microorganisms to consume. All other things being equal, the overproducing microorganism is then at a competitive disadvantage, and usually it fails to survive. All other things are not always equal in nature, however, and therefore some microorganisms do manage to survive in their ecological niche even though their metabolism is somewhat less regulated than that of other microorganisms. The less regulated strains are much sought after by microbiologists in large-scale screening programs. Once a slightly deregulated microorganism is brought into the laboratory the microbiologist attempts to exploit or bypass the natural regulatory controls by manipulating either the nutrition or the genetics of the culture.

The most important primary metabolites in the fermentation industry are amino acids, purine nucleotides, vitamins and organic acids. Citric acid, for example, is made by molds under conditions where a nutrient imbalance is created by limiting the supply of certain minerals such as iron and manganese. In most industrial processes genetic and environmental manipulations are combined to achieve remarkable levels of metabolite formation. Notable examples are the 20,000-fold overproduction of riboflavin (vitamin B_2) by the mold *Ashbya gossypii* and the 50,000-fold overproduction of cobalamin (vitamin B_{12}) by the bacteria *Propionibacterium shermanii* and *Pseudomonas denitrificans*.

Of all the traditional products made by fermentation the most important to human health are the secondary metabolites. This group includes not only antibiotics but also toxins, alkaloids and plant growth factors. They vary widely in structure, are each manufactured by only one microbial species or a small number of species and are often formed as a mixture of closely related substances. In nature their functions serve the survival of the species, but when the microorganisms producing them are grown in pure culture, the secondary metabolites have no such role.

The best-known secondary metabolites are the antibiotics. More than 5,000 antibiotics have already been discovered, and new ones are still being found at a rate of about 300 per year. Most are useless: they are either toxic or inactive in living organisms. For some unknown reason the actinomycetes are amazingly prolific in the number of antibiotics they can secrete. Roughly 75 percent of all antibiotics are obtained from these filamentous prokaryotes, and 75 percent of those are in turn made by a single genus: *Streptomyces*.

In spite of the large number of known antibiotics, the search for new ones continues. New antibiotics are needed to combat both naturally resistant organisms and organisms that have acquired resistance through mutation; they are also needed to provide safer drugs. Chemists work steadily to modify the natural structures uncovered by microbiologists. Such semisynthetic antibiotics are already important in clinical practice. Antibiotics also serve purposes other than human and animal chemotherapy, such as the promotion of growth in farm animals and the protection of plants against inimical microorganisms.

The most important factor in keeping the industrial-fermentation industry productive and competitive with the nonbiological chemical industry is mutation. The industrial microbiologist can treat an organism with a mutagenic agent that increases the frequency of changes in the genes of the cells by several orders of magnitude. Although the genetic changes that take place are usually detrimental to the organism, they are sometimes beneficial to man. The microbiologist can often identify these changes (for example an increase in antibiotic production) by appropriate screening procedures and can preserve them indefinitely. As a result today's industrial strains of *Penicillium chrysogenum* can produce 10,000 times more penicillin per unit volume of broth than Fleming's original culture did.

In addition mutants occasionally manufacture a modified antibiotic with improved properties. Although this outcome is usually a matter of chance, microbiologists have recently developed a technique called mutational biosynthesis by which new antibiotics can be developed in a more rational way.

Although the record of the mutagenic approach in industry has been a good one, it is nonetheless a slow and painstaking procedure. In recent years developments in microbial genetics have been so rapid and dramatic that they have created an entirely new set of options for the fermentation industry. These options include protoplast fusion, gene amplification and recombinant-DNA technology. In nature genetic changes can arise not only through mutation but also through genetic recombination between two cells of different genetic types, yielding progeny with genes from both parents. Until recently this phenomenon had not been exploited much in industry because of the extremely low frequency of genetic recombination in industrial strains of microbial cells. For example, crossing two strains of the same species of *Streptomyces* leads to only one recombinant cell among a million nonrecombinant cells.

In the new technique of protoplast fusion, however, the cell walls of each type are removed, the resulting protoplasts are mixed and the fusion product is allowed to regenerate its cell wall. This procedure leads to a remarkable increase in the frequency of genetic recombination, with the result that many species of *Streptomyces*, after crossing, give rise to as many as one recombinant cell in every five cells. The increase in the frequency of recombination also makes it possible to detect genetic recombination between different species. Protoplast fusion is now being exploited to recombine slow-growing but high-producing mutants with their fast-growing ancestors to yield fast-growing and high-producing strains, to recombine several high-producing mutants from a single mutagenic treatment or separate treatments to yield an additive, or perhaps even a synergistic, combination of improved production mutations, and to produce new hybrid antibiotics by mating closely related strains that make different antibiotics.

Gene amplification is another approach to genetic manipulation with a great potential in industrial microbiology. In this technique genes are amplified, or duplicated, by forcing plasmids that carry them to reproduce rapidly. (Plasmids are small circular pieces of extrachromosomal DNA, carrying as few as two genes or as many as 250, that can exist autonomously in the cytoplasm of a cell or as an integral part of the chromosome.) When the plasmids are present in the autonomous state, they usually reproduce at the same rate as or at a somewhat higher rate than chromosomes. Although there are normally between two and 30 copies of a plasmid per cell, the plasmids can be forced into reproducing much faster than chromosomal DNA, yielding as many as 3,000 copies of a plasmid per cell.

The technique of gene amplification has been widely exploited in bacteria such as *Escherichia coli*. It is now possible in principle to transfer any chromosomal gene (or cluster of genes) to a plasmid and to amplify the gene, increasing the manufacture of the protein for which the gene codes to very high levels. Almost all bacterial species contain plasmids, as do eukaryotes such as yeasts. Of great importance to the fermentation industry is the fact that virtually all antibiotic-producing species contain plasmids that incorporate either structural genes for the manufacture of antibiotics or genes that regulate the expression of such structural genes.

Not only plasmids but also bacteriophages can serve for transferring and amplifying genes. New processes for the manufacture of enzymes and primary metabolites will probably result from the introduction of gene-amplification techniques, for the reason that many of the enzymes coding for the structural genes of primary-metabolite biosynthesis are clustered on the chromosomes of

ACTIVATED-SLUDGE PROCESS employs a complex population of microorganisms for the detoxification and degradation of sewage and industrial wastes. A large metropolitan sewage-treatment plant based on the process is seen in this vertical aerial photograph, made from an altitude of about a mile. The plant is southwest of Chicago, on the bank of the Chicago Sanitary and Ship Canal. North is to the left.

bacteria. The transfer of these "operons" to the DNA of plasmids or bacteriophages, followed by gene amplification, could yield effective new industrial microorganisms.

The highly publicized achievements of the artificial recombination of DNA will clearly have a great impact on industrial microbiology in the next decade. Genetic recombination is a means of increasing the diversity of microorganisms; it is the bringing together of genetic information to form new stable combinations and thus create new genotypes. In nature genetic recombination occurs between organisms of the same species or closely related species. All organisms have enzymes known as restriction endonucleases that recognize foreign DNA and destroy it so that "illegitimate" recombination does not occur. In 1973 it was discovered that it is possible to cut DNA molecules with restriction enzymes, to join pieces of DNA with another enzyme (DNA ligase) and to reintroduce the recombinant DNA into E. coli with the aid of a plasmid as a vector. Soon afterward plasmid genes from unrelated bacterial species were recombined in the test tube and expressed in E. coli. Later it was found that DNA from a eukaryote (a yeast) could express itself in a bacterium. Since then many mammalian genes, including human ones, have been cloned: copied in vast numbers by the multiplication of the bacterium into which they were introduced. Initial concerns about the safety of these procedures appear to be unwarranted in the light of additional evidence.

Today bacteria are making many proteins they never made in nature; the best-known are insulin and interferon. Insulin produced by bacteria has been shown to be active and safe in human beings, and interferon produced by recombinant-DNA technology is currently being tested in patients. From the example of interferon it is easy to appreciate the ability of recombinant-DNA technology to make protein molecules available at a reasonable cost: before interferon was made in bacteria 50 milligrams of very impure material cost some $2 million; at some point in the future bacterially produced interferon could cost as little as pennies per milligram of pure material. Recombinant DNA will be exploited to make hormones, analgesics and vaccines, and these products will be purer than those supplied by earlier technologies. Traditional fermentation processes, particularly those of the enzyme industry, will be markedly improved by the application of recombinant-DNA technology. Moreover, the new technology may well phase out many energy-intensive and highly polluting chemical-manufacturing processes. It is already being applied to the development of new bacterial strains that will convert agricultural and forest biomass into liquid fuel and chemicals.

The study of genetic recombination in higher organisms has also led to a major development in the field of immunology. In 1975 Georges Köhler and Cesar Milstein of the British Medical Research Council Laboratory of Molecular Biology in Cambridge fused a myeloma, or skin-cancer cell, of a mouse with an antibody-producing white cell to make a "hybridoma," or hybrid cell, that grew in the test tube and manufactured a pure specific antibody. Never before had such pure monoclonal antibodies been made; investigators and clinicians had to rely on impure mixtures of antibodies in animal serum to provide immunological protection against disease. Today monoclonal antibodies are commercially available for a variety of purposes.

Industrial microbiology is exciting in still other arenas. For example, the physiology of the cells of higher organisms is being intensively examined with a view toward exploiting such cells to manufacture plant and animal metabolites. Some secondary metabolites have been made with cultures of plant cells, but the technology is not yet commercially feasible. Human interferon and other mammalian proteins have been produced with cultures of animal cells growing on microscopic beads (called microcarriers). Such cultures make it possible to greatly increase the ratio of surface area to volume of fluid for cells that must be attached to a surface in order to grow. Attempts to commercialize microcarrier cultures are already under way.

Major agricultural advances, such as the replacement of synthetic nitrogen fertilizers by enhancing the efficacy of natural nitrogen-fixing microorganisms, should come from the new concepts of genetic engineering. One avenue where progress is being made is the establishment of a synergistic relation between free-living nitrogen-fixing bacteria and plants such as corn. In this approach ammonia-excreting strains of *Azotobacter vinelandii* provide fixed nitrogen to the plants, and the plants supply carbon to the bacteria.

A new pharmaceutical approach is the application of secondary metabolites to diseases that are not caused by bacteria or fungi. For years the major drugs available for the treatment of noninfectious diseases have been strictly synthetic products prepared by chemists. Similarly, major therapeutic agents for the treatment of parasitic diseases in animals have been obtained by the screening of chemically synthesized compounds and the modification of their molecular structure. Although thousands of compounds have been tested, only a few promising ones have been uncovered. As new compounds of this type become harder to find, microbially produced substitutes are filling the void.

Another application of microbial activity is the detoxification and degradation of sewage and industrial waste. The usefulness of microorganisms in waste treatment has been recognized since 1914, when the activated-sludge process was first developed. The sludge process depends on a complex population of microorganisms that form naturally because of each organism's ability to degrade a constituent of the waste material and to coexist with the others in a nutritionally complementary system. The next advance was to enrich the sludge by inoculating it with the desired mixture of microorganisms. Now pure cultures of a single microorganism are being made available to degrade specific compounds in industrial waste, such as polychlorinated biphenyls (PCB's).

Oil spills and the release of ballast and wash waters from oil tankers are other waste problems microbiology may be able to solve. Many microorganisms that can consume components of petroleum have been isolated. Strains of one such bacterial species, *Pseudomonas putida,* carry plasmid genes coding for enzymes that can degrade different components of oil. By genetic engineering the capacity to degrade the various components has been combined in one strain. Such a multiplasmid strain degrades petroleum faster than any of the original strains.

Certain microorganisms are the basis of a metallurgical process that is thought to go back to the Romans: the bacterial leaching of low-grade ores to extract metals from them. Today copper and uranium are commercially leached by bacteria, mainly members of the genus *Thiobacillus.* New approaches to such bacterial leaching are also being made. They include the examination of acid- and heat-resistant microorganisms (including fungi) for their extraction abilities, the investigation of the mechanisms underlying the affinity of bacteria for metals and the genetic manipulation of bacteria to increase their resistance to the toxicity of the metals.

In reviewing the history and current state of industrial microbiology we are struck by an abiding theme: mutually beneficial relations between what we have come to call basic research and applied research. A century ago the largely practical investigations of Pasteur led to the establishment of microbiology, immunology and biochemistry. Much later the discovery of antibiotics by applied microbiologists provided tools crucial to the development of molecular biology. And now basic research in microbial genetics has returned the favor by supplying an array of new techniques for industrial applications. This synergy between science and technology, we believe, is the key to further progress in industrial microbiology.

BACTERIA ARE EXPLOITED on a vast scale in extracting certain metals from low-grade ores. These two photographs, for example, show bacterial-leaching operations at two large open-pit copper mines in the southwestern U.S. The aerial photograph at the top gives an overview of an entire mining site at Santa Rita, N.M.; the mine itself is in the background, and the associated leaching dumps are in the right foreground. The photograph at bottom is a closeup view of a leaching dump at a similar mining site at Bingham Canyon, Utah; the leaching solution can be seen being recycled by an array of rotating sprinklers. The bacteria, mainly members of the genus *Thioba-* *cillus,* assist in the leaching operation by converting iron in various compounds from the ferrous form into the ferric. The ferric iron, an effective oxidizing agent, then performs two useful functions: it oxidizes pyrite to form sulfuric acid, thereby maintaining the high acidity of the leaching solution, and it oxidizes insoluble copper-containing sulfide minerals to produce soluble copper sulfate, which migrates in solution to the bottom of the dump, where it collects in catch basins. The solution is periodically pumped out to facilities where the copper is recovered. Huge numbers of bacteria are involved in the leaching operation: in places more than a million per gram of ore.

2

Industrial Microorganisms

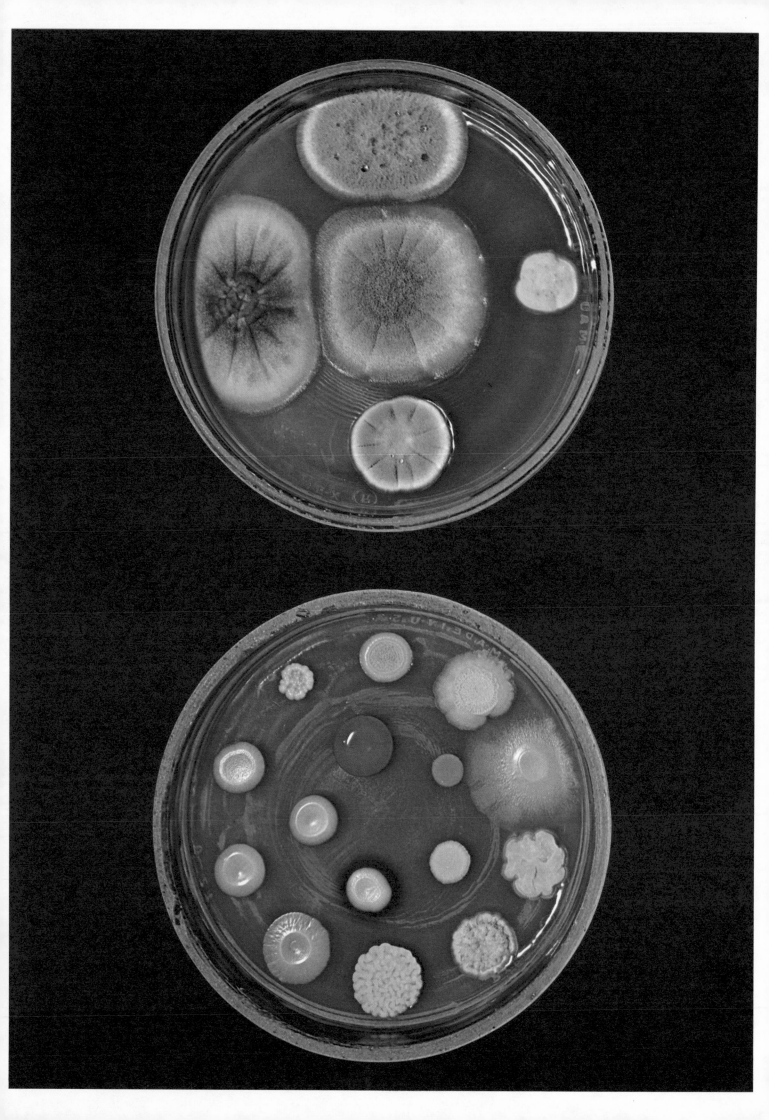

Industrial Microorganisms

BY HERMAN J. PHAFF

They are yeasts, molds, bacteria and actinomycetes (filamentous bacteria). They now include, however, cultured mammalian cells and "hybridomas": cells created by the fusion of two cell lines

The microorganisms that make products useful to man represent at most a few hundred species out of more than 100,000 that exist in nature. Their utility in brewing, wine making and leavening bread was discovered quite by accident. The yeasts that transform grain mash, grape juice and bread dough are ubiquitous organisms, as are the bacteria that sour milk and the molds that impart the distinctive character of diverse kinds of cheese. To these three groups of microorganisms with industrial applications—yeasts, molds and bacteria—must be added a fourth group, the soil-inhabiting actinomycetes, filamentous bacteria whose value as a source of antibiotics has been recognized only since the 1940's. And to all of these must now be added cells that do not live free in nature but that too can manufacture substances useful in the diagnosis and treatment of disease: mammalian cells grown in culture.

There is no ready way to classify microorganisms into the useful and the nonuseful. All are useful in the sense that they help to recycle the molecules of the organic world. In this role they are not merely useful but indispensable. A considerable number of microorganisms can of course be harmful to animals and plants. The large majority, however, are normally innocuous. The few that have been found to be industrially useful are prized simply because they happen to elaborate a substance that is recognized to have value and that is not as readily or as cheaply obtainable in any other way. In a few instances microbial cells are cultivated for their own sake, for example in the production of baker's yeast. More often the desired substance is a metabolic product, such as ethanol.

Most bacteria are small unicellular organisms, measuring only one micrometer (a millionth of a meter) or a few micrometers. Most yeasts are also unicellular, but they are larger, usually from six to 12 micrometers. Molds, on the other hand, are multicellular, and whereas their individual cells are small (rarely more than 25 micrometers in their longest dimension) the overall fungal body is readily visible to the unaided eye. The sexually reproducing fungi can have fruiting bodies (such as mushrooms and truffles) that are quite sizable, consisting of billions of cells.

Microorganisms are normally divided into two large groups: prokaryotes and eukaryotes. The prokaryotic microorganisms, regarded as the more primitive of the two, have a single circular chromosome of double-strand DNA that is unconfined within the cell's cytoplasm. The eukaryotic microorganisms, which are much larger than the prokaryotes, have at least two chromosomes (and in some species more than 20) enclosed in a nuclear envelope with a porous double membrane. The chromosomes of eukaryotes are linear and are intimately associated with the class of proteins named histones. Bacteria are prokaryotes; yeasts and fungi are eukaryotes.

For their growth and multiplication microorganisms conduct a wide variety of metabolic processes to obtain energy and new cell material. A few microorganisms that are photosynthetic are able to use the energy of light to convert carbon dioxide from the air, along with hydrogen from water, into cellular organic material, as is done by the higher plants. None of the common industrial microorganisms, however, is photosynthetic. One exception, in which interest has largely lapsed, is certain species of algae that a few years ago were considered promising as a source of food protein. The common industrial microorganisms require organic substrates for growth.

Microorganisms can be divided by their environmental requirements into

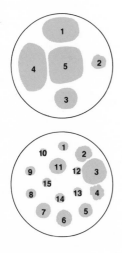

MOLDS
1 *Penicillium chrysogenum*
2 *Monascus purpurea*
3 *Penicillium notatum*
4 *Aspergillus niger*
5 *Aspergillus oryzae*

YEASTS
1 *Saccharomyces cerevisiae*
2 *Candida utilis*
3 *Aureobasidium pullulans*
4 *Trichosporon cutaneum*
5 *Saccharomycopsis capsularis*
6 *Saccharomycopsis lipolytica*
7 *Hanseniaspora guilliermondii*
8 *Hansenula capsulata*
9 *Saccharomyces carlsbergensis*
10 *Saccharomyces rouxii*
11 *Rhodotorula rubra*
12 *Phaffia rhodozyma*
13 *Cryptococcus laurentii*
14 *Metschnikowia pulcherrima*
15 *Rhodotorula pallida*

MOLDS AND YEASTS are microorganisms that form visible and often colorful structures when they alight or are deposited on a suitable medium. In the photograph on the opposite page pure cultures of several molds (*top*) and yeasts (*bottom*) are shown eight to 10 days after they were seeded on a nutrient agar in glass dishes. The maps at the left identify the five molds and the 15 yeasts. Four of the molds (*1, 3, 4, 5*) and five of the yeasts (*1, 2, 6, 9, 10*) yield commercially useful products, including beer, citric acid, enzymes, antibiotics, sake, soy sauce and microbial protein (*see illustration on page 88*). One of the yeasts, *Phaffia rhodozyma* (*12*), is being tested as a food supplement for hatchery-raised fish, the flesh of which tends to be white. The yeast synthesizes a carotenoid, astaxanthin, which turns the flesh to the normal orange pink.

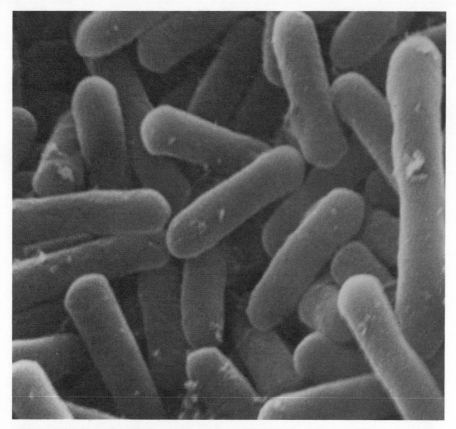

PROKARYOTIC INDUSTRIAL MICROORGANISMS are here represented by numerous bacteria of the species *Bacillus brevis*, which manufactures the antibiotic gramicidin *S*. This scanning electron micrograph, which enlarges bacteria 9,500 diameters, was made by Erika A. Hartwieg of the Electron Microscopy Facility at the Massachusetts Institute of Technology.

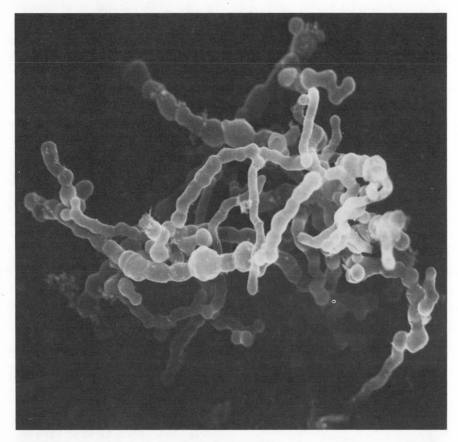

EUKARYOTIC INDUSTRIAL MICROORGANISMS are here represented by the hyphae, or filaments, of the mold *Cephalosporium acremonium*, which manufactures the antibiotic cephalosporin. These swollen filaments coincide with the high-production phase of the organism. The micrograph, which enlarges the hyphae 700 diameters, was also made by Hartwieg.

three groups. In one group are the strict aerobes, which can metabolize and grow only in the presence of atmospheric oxygen. In the second group are the strict anaerobes, which not only metabolize and grow in the absence of free oxygen but also require its exclusion lest they be harmed by it. In the third group are the facultative organisms, which are able to switch their metabolic machinery from an aerobic (respiratory) mode to an anaerobic (fermentative) one, depending on the environment in which they find themselves. Among the strict aerobes are the prokaryotic streptomycetes (a source of antibiotics) and most filamentous eukaryotic fungi (associated with cheeses and fermented foods). Strict anaerobes are represented by members of the bacterial genus *Clostridium*, of which *C. botulinum*, the source of the toxin of botulism, is a notorious example. Industrial yeasts, which can either respire or ferment certain substrates, are facultative organisms.

Anaerobic metabolism is always less efficient than respiration because fermentation does not exploit all the energy in the organic substrate (for example sugar) for making the universal fuel of the cell (the compound adenosine triphosphate, or ATP) and ultimately for making the substance of the cell. Some potential substrate is excreted from the cells in the form of degradation products that could be further oxidized into carbon dioxide and water. The products of fermentation, such as the ethanol released by yeast fermentation, cannot be metabolized further under anaerobic conditions by the organism that manufactures them.

The types of fermentative biochemical pathways leading to useful (hence incompletely metabolized) products are quite variable. For example, yeasts can ferment one molecule of a six-carbon monosaccharide sugar such as glucose or fructose into two molecules of ethanol and two molecules of carbon dioxide. Pathways can also be of two general types: homofermentative (signifying one principal product) or heterofermentative (two or more products). One group of lactic acid bacteria are homofermentative: they convert glucose into lactic acid. Another group of the same bacteria are heterofermentative: they convert glucose by a different biochemical pathway into lactic acid, ethanol and carbon dioxide. *Clostridium acetobutylicum* is another heterofermentative organism. It converts glucose into a mixture of acetone, ethanol, isopropanol and butanol.

Aerobic growth, on the other hand, enables some organisms to completely oxidize a certain fraction of the substrate and thereby extract a maximum amount of energy for converting the remaining substrate into cell mass. If the purpose of the industrial fermentation is to maximize cell mass, as in the produc-

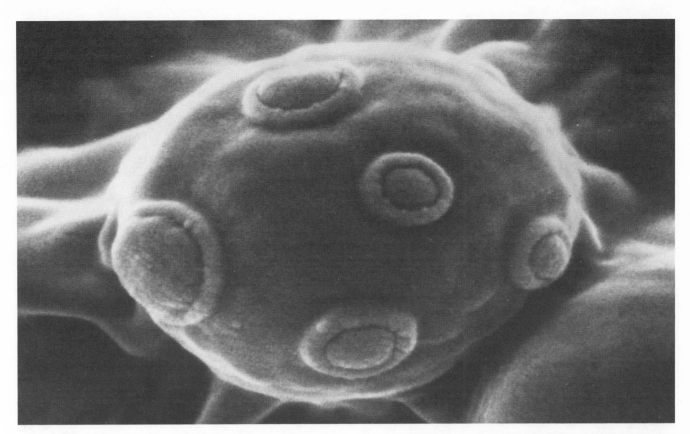

YEAST CELL, which is also a eukaryote, appears in this scanning electron micrograph of the species *Saccharomyces cerevisiae*. Yeasts can reproduce asexually by budding or sexually (*see top illustration* *on page 19*). The protuberances on this cell are scars where cells budded off. The micrograph, which enlarges the cell 12,500 diameters, was made by Martin W. Miller of University of California at Davis.

tion of baker's yeast or of microbial protein as a food source, it is clearly advantageous to have aerobic growth with complete utilization of the substrate by respiration. It may be asked, however, how aerobic growth can lead to the manufacture of useful microbial products if all the substrate not converted into cell mass is oxidized into carbon dioxide and water.

One answer is that not all oxidative reactions catalyzed by strictly aerobic microorganisms go to completion. An example is the conversion of ethanol into acetic acid (vinegar) by acetic acid bacteria: CH_3CH_2OH (ethanol) $+ O_2$ $\rightarrow CH_3COOH$ (acetic acid) $+ H_2O$. The acetic acid bacteria are also capable of incomplete oxidations on other substrates, such as the conversion of glucose into gluconic acid. Although such "underoxidizers" can derive energy in the form of ATP from limited oxidations, they generally cannot derive carbon "skeletons" for growth from incompletely oxidized substrates and therefore are dependent for growth on other nutrients supplied in the medium.

In other instances useful organic compounds can be elicited from aerobic organisms by deliberately manipulating the biosynthetic pathways by which the organism converts the substrate into the thousands of different molecules that constitute a living cell. In normal metabolism each compound the cell needs is usually made in just the right amount. This is accomplished by a series of strict regulatory reactions that halt the manufacture of the intermediates and the end products of a metabolic pathway when a particular compound reaches a certain concentration. The industrial microbiologist has learned to select mutant strains in which this exquisite regulatory process is crippled in a desirable way. For example, in normal cells the synthesis of lysine, one of the 20 amino acids from which all cellular proteins are made, is regulated so that only the amount needed for the cell's thousands of different proteins is made. Certain mutants of *Corynebacterium glutamicum* have been found, however, in which the lysine regulatory mechanism is so defective that lysine is overproduced to the extent of more than 50 grams per liter of nutrient medium. Lysine and similar products of low molecular weight that are essential components in cell growth are called primary metabolites.

Another group of industrially important microbial products, called secondary metabolites, are compounds not required for cellular biosynthesis. Such products are synthesized by certain microorganisms, usually late in the growth cycle, for reasons that are often obscure. The best-known examples are antibiotics. Since secondary metabolites play no direct role in the energy metabolism and growth of the organism, they presumably contribute to the organism's survival by inhibiting competitors that could otherwise occupy the same ecological niche.

Organisms that secrete secondary metabolites initially go through a period of rapid growth, the trophophase, in which the synthesis of the secondary substance is negligible. When further growth is halted by the depletion of one or more essential nutrients in the medium, the organism enters the idiophase: the phase peculiar to that organism. The trigger for the synthesis of secondary metabolites in the idiophase is not known. Most organisms are sensitive to their own antibiotics during the trophophase but become physiologically resistant during the idiophase, so that the delay in the secretion of secondary metabolites is obviously crucial to keeping antibiotic-producing organisms from destroying themselves.

Still a third class of industrially important substances synthesized by microorganisms are the proteins that act as enzymes. Since typical proteins consist of several hundred amino acid units, few have ever been synthesized in the laboratory and no natural enzyme has ever been synthesized industrially. Microorganisms rely on catabolic enzymes to degrade complex substrates into simpler molecules that can then be assimilated. Anabolic, or biosynthetic, enzymes carry out the step-by-step reactions that rebuild the simple molecules into the substances (including enzymes of both types) needed for cell metabolism and growth. As with amino acids, the cell normally synthesizes only as much of each enzyme as it needs. As with amino acid regulation, however, organisms can be selected that overproduce particular enzymes when they have the right environment and the appropriate nutrients.

One method of increasing enzyme synthesis is induction. The genetic blueprint for each enzyme is the sequence of DNA nucleotides termed a structural gene, residing either in the single chromosome of a prokaryote or in one of the several chromosomes of a eukaryote. The structural genes that code for the synthesis of many enzymes are normally inactive in the absence of the enzyme substrate or a molecule structurally analogous to it. Enzyme production is then said to be repressed. When the required substrate or analogue is added to the medium, the structural gene is activated and the enzyme is synthesized. Such an event is called derepression, or induction, and the enzymes that respond are called inducible enzymes (to distinguish them from constitutive enzymes, which are not affected in this way).

In some instances the inducer is the product of an enzymatic reaction. For example, the sugar maltose (actually an intermediate in the metabolism of the sugar) can induce the fungus *Aspergillus niger* to begin synthesizing glucamylase, an enzyme that breaks down the chain of sugars in starch into glucose. Although the substrate on which glucamylase acts is starch, starch does not have to be present in the medium in order to induce the synthesis of the enzyme. It turns out that some analogues that are poor or inactive substrates can be extremely potent inducers.

The most widely held model of induction is the one devised some 20 years ago by François Jacob and Jacques Monod of the Pasteur Institute. Briefly, in cells not exposed to an inducer the structural gene for the enzyme cannot be transcribed into messenger RNA (the first step in translating a gene into an enzyme) because RNA polymerase, the enzyme needed to carry out the synthesis of messenger RNA, is kept from acting by virtue of the fact that an operator gene adjacent to the structural gene on the DNA is blocked by a "repressor" protein. The repressor protein in turn is coded for by a nearby repressor gene. When inducer molecules are present, they combine with repressor molecules and so can no longer combine with the operator gene. With the operator gene free to function the RNA polymerase can transcribe the structural gene into messenger RNA, which then directs the assembly of the enzyme from the appropriate amino acids.

Certain catabolic enzymes of industrial importance, such as the amylases (starch digesters) and the proteases (protein digesters), can be obtained from microorganisms in supernormal amounts by circumventing the phenomenon known as catabolite repression. The term describes the decrease in the rate of synthesis of a catabolic enzyme when the microbial cells are exposed to a source of carbon that is rapidly assimilated. Because such a carbon source is often glucose the phenomenon is sometimes known as the glucose effect. In a few instances even the rapid catabolism of the inducer itself has been observed to give rise to catabolite repression. In such instances a slow feeding of the actual inducer (or the substitution of a slowly metabolized analogue inducer) can stimulate large increases in the synthesis of the desired enzyme.

The underlying cause of catabolite repression is that in cells that are provided with a rapidly utilizable carbon source there is a sharp decrease in the intracellular concentration of cyclic $3',5'$-adenosine monophosphate (cAMP). In the absence of sufficient cAMP the structural genes coding for the particular enzymes are not transcribed effectively, and little or no enzyme is synthesized. An understanding of catabolite repression is of great importance in industrial microbiology because many commercially valuable enzymes are obtained from microorganisms subject to this phenomenon.

An altogether different type of metabolic product harvested in large volume from microorganisms is capsular polysaccharides, notably dextran and xanthan gum. Dextran, a large glucose polymer with a molecular weight of between 50,000 and 100,000 daltons, can serve as an extender for blood plasma. When its chains are cross-linked, it yields beads that are effective as molecular sieves. Xanthan gum, which has been found safe for human consumption, is added to many food products as a thickening agent and stabilizer. It also finds use in such diverse fields as textile printing and dyeing, oil-well drilling (as an additive to drilling mud) and the formulation of cosmetics and pharmaceuticals.

Capsular polysaccharides are substances of high molecular weight that form a thick capsule around the cells of certain microorganisms. Dextran is made in large amounts by *Leuconostoc mesenteroides* and related lactic acid bacteria, but only when they are grown on sucrose as the substrate. In the bacterial capsule dextran is a glucose polymer with a molecular weight varying from 15,000 daltons to 20 million, depending on the strain of bacterium. The bacteria harbor an enzyme called either transglu-

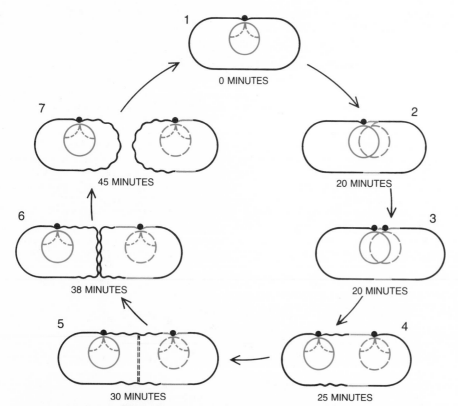

REPRODUCTION OF BACTERIA is accomplished asexually by cell division. The diagram shows seven stages in the life cycle of the colon bacterium *Escherichia coli*, which has a doubling time of 45 minutes. In the newborn cell (*1*) the single circular chromosome of DNA (*color*) is already being replicated (*broken lines*). After 20 minutes the new chromosome is complete and is affixed to an attachment site within the cell (*2, 3*). By 25 minutes the two chromosomes have begun to replicate (*4*) and a septum, or dividing membrane, appears in the middle of the cell (*5*). By 38 minutes the septum is a wall (*6*). Seven minutes later division is complete (*7*).

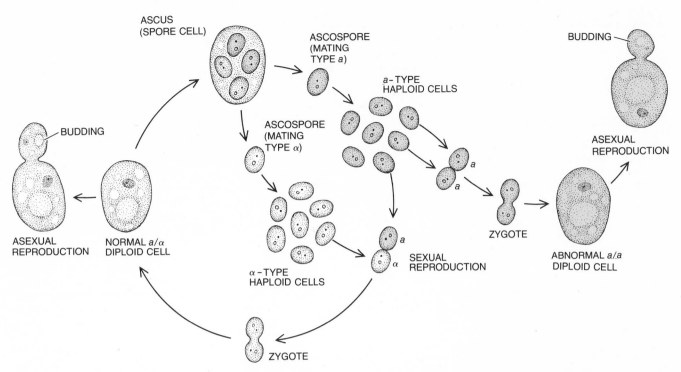

REPRODUCTION OF YEAST is normally asexual, proceeding by the formation of buds on the cell surface, but sexual reproduction can be induced under special conditions. In the sexual cycle a normal diploid cell (a cell with two sets of chromosomes and therefore two sets of genes) gives rise to asci, or spore cells, that contain four haploid ascospores (cells with one set of chromosomes and one set of genes).

The ascospores are of two mating types: a and α. Each type can develop by budding into other haploid cells. The mating of an a haploid cell and an α haploid cell yields a normal a/α diploid cell. Haploid cells of the same sex can also unite occasionally to form abnormal diploid cells (a/a or α/α) that can reproduce only asexually, by budding in the usual way. Industrial yeasts reproduce mainly by budding.

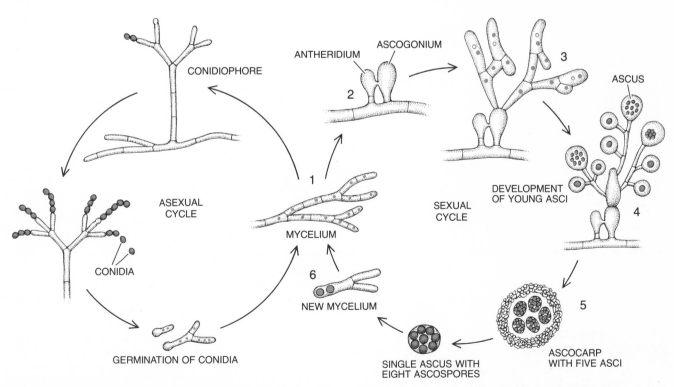

REPRODUCTION OF A MULTICELLULAR FUNGUS, such as one of the higher Ascomycetes, can be asexual or sexual. The details vary with genus and species. The branched vegetative structure common to both reproductive cycles is the mycelium, composed of hyphae (1). In the asexual cycle the mycelium gives rise to conidiophores that bear the spores called conidia, which are dispersed by the wind. In the sexual cycle the mycelium develops gametangial structures (2), each consisting of an antheridium (containing "+" nuclei) and an ascogonium (containing "−" nuclei). The nuclei pair in the ascogonium but do not fuse. Ascogenous, binucleate hyphae develop from the fertilized ascogonium (3), and the pairs of nuclei undergo mitosis, which replicates the newly paired chromosomes. Finally nuclei fuse in the process called karyogamy (4) at the tips of ascogenous hyphae. That is the only diploid stage in the life cycle. Soon afterward the diploid nuclei (*large colored dots*) undergo meiosis, or reduction division. The result is eight haploid nuclei (*small colored dots*), each of which develops into an ascospore. At the same time the developing asci are enclosed by mycelial hyphae in an ascocarp (5). In the example shown here the ascocarp is a cleistothecium, a closed structure. Ascospores germinate to yield binucleate or multinucleate mycelium (6).

cosidase or dextransucrase that splits the disaccharide sucrose into fructose and glucose. Fructose provides for the bacteria's growth; glucose is transferred molecule by molecule to a growing strand of dextran. The enzymatic synthesis of dextran proceeds either with whole cells or with a cell-free extract. Xanthan gum is synthesized by the bacterium *Xanthomonas campestris* when it is grown aerobically on glucose media. The polysaccharide made by this organism is branched and more complex than dextran. It is assembled from glucose, mannose (also a six-carbon monosaccharide) and glucuronic acid, some of which have acetyl (CH_3CO) and pyruvate (CH_3COCO) groups attached to their molecule.

Yeasts were exploited for thousands of years in the making of alcoholic beverages and for leavening bread before yeast cells were recognized as microorganisms and the true nature of fermentation was discovered. The presence of yeast cells in fermenting beer was first recorded in 1680 by Anton van Leeu-

wenhoek, who is credited with inventing the forerunner of the modern microscope. Nearly 200 years later, in 1876, Louis Pasteur presented his views on fermentation in the classic work *Études sur la bière,* in which he postulated that microorganisms living under anaerobic conditions are able to live and to grow by substituting the process of fermentation for the better-understood respiratory process of many organisms. He recognized that the fermentation process that converts sugars into alcohol and carbon dioxide supplies the energy necessary for yeast cells to live in the absence of oxygen. Pasteur further recognized that when oxygen is available to yeast cells, fermentation is repressed and is supplanted in varying degrees by respiration. In the latter process the sugar is fully oxidized to carbon dioxide and water.

Whereas the enzymes of fermentation are constitutive the enzymes of respiration are inducible. Fermentation enzymes reside in the cytoplasm of the cell; the respiratory enzymes are in the

organelles called mitochondria. The respiratory enzymes are subject to catabolite repression by glucose. For that reason when yeast is grown with a plentiful supply of air in order to maximize the cell mass, as in the production of baker's yeast, it is essential that the cell's nutrient sugar solution be fed at a rate such that it never exceeds a few tenths of 1 percent. In this way the catabolite repression of the respiratory enzymes is prevented and practically all the added sugar is respired rather than converted into alcohol by fermentation.

The only sugars that can be fermented by yeasts are the six-carbon monosaccharides (or polymers of such sugars). Disaccharides such as sucrose and maltose are first broken down by the cell's hydrolytic enzymes to monosaccharides. In industrial processes such as brewing and whiskey making, where starches serve as the main substrate for fermentation, the starches are usually broken down to monosaccharides by the addition of amylase enzymes from barley malt or from certain species of molds. Although there are yeast species that can grow on starch by virtue of making their own amylases, such species are not efficient enough for industrial alcohol production.

When yeasts are grown in the respiratory mode, a much broader range of compounds can be exploited as substrates, depending on the yeast species selected. Some yeasts metabolize few compounds, others can handle many. For example, the ability of *Candida utilis* (a species employed for food yeast) to metabolize the pentose (five-carbon) monosaccharides xylose and arabinose makes this species suitable for growing on sulfite waste liquor, a by-product of the paper industry.

Other yeasts, such as *Saccharomycopsis lipolytica,* are able to metabolize straight-chain hydrocarbons that range from 10 to 16 carbon atoms in length. In pilot installations yeasts have been grown on purified fractions of petroleum. The yeast cells first oxidize the hydrocarbons to long-chain fatty acids by means of hydroxylation enzymes as an intermediate step. The fatty acids are broken down by a special oxidation process that yields acetyl-coenzyme A, which is ultimately transformed into cell material. Another substrate of industrial interest is methanol (CH_3OH), a simple alcohol that can be made by the oxidation of the gas methane or derived from coal. Methanol can be assimilated by a limited number of yeast species by virtue of a novel metabolic process involving specialized cell organelles called microbodies. Yeast grown in this way can serve as a protein supplement in animal feeds.

Nearly all yeasts are capable of converting inorganic nitrogen into proteins and nucleic acids. The nitrogen can be

ACETONE-BUTANOL FERMENTATION is carried out anaerobically by the bacterium *Clostridium acetobutylicum.* **The diagram shows the series of intracellular reactions that convert two molecules of glucose into one molecule of butanol, one molecule of acetone, four molecules of hydrogen and five of carbon dioxide. Small amounts of ethanol are formed as a minor product. In the process four molecules of adenosine diphosphate (ADP) and four units of inorganic phosphate (P_i) are converted into four molecules of adenosine triphosphate (ATP), the universal fuel of intracellular processes. Acetyl-CoA (acetyl-coenzyme A) is a coenzyme that plays a key role in the metabolism of all cells. The molecule nicotinamide adenine dinucleotide (NAD) is an acceptor of hydrogen atoms; the reduced form of the molecule, $NADH_2$, is a donor of hydrogen atoms. Specific enzymes carry out the reactions indicated by asterisks.**

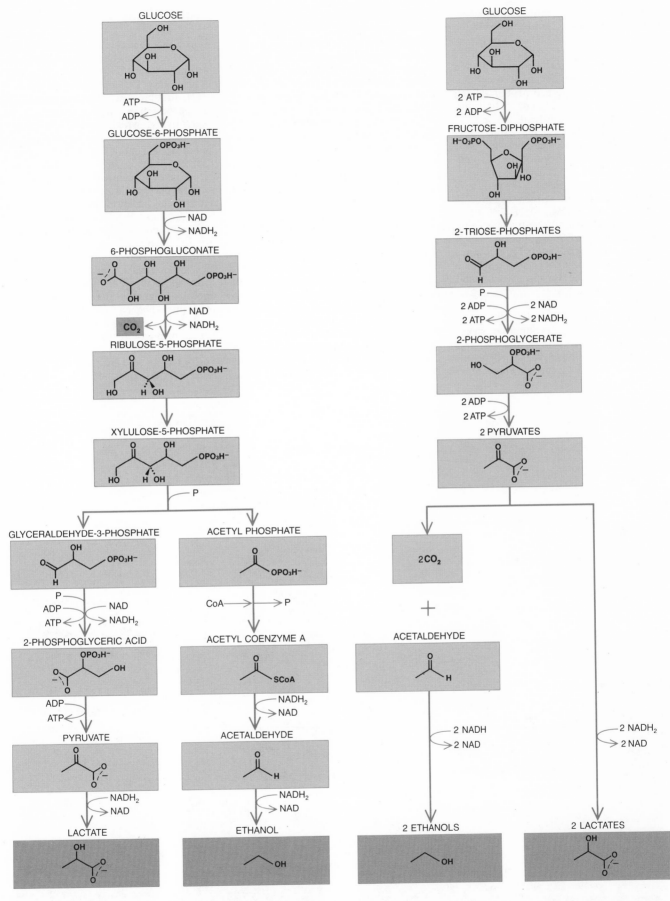

HETEROFERMENTATIVE AND HOMOFERMENTATIVE pathways are compared in two reaction sequences. The diagram at the left shows the scheme by which the heterofermentative lactic acid bacterium *Leuconostoc* transforms one molecule of glucose into one molecule each of carbon dioxide, lactic acid and ethanol. The diagram at the right shows how two homofermentative organisms, a yeast and a bacterium, ferment glucose with the same sequence of reactions up to the point where two molecules of pyruvic acid have been formed. The pathways thereafter diverge. The yeast, *Saccharomyces cerevisiae*, converts the two pyruvic acid molecules into two molecules each of carbon dioxide and ethanol. The bacterium, a streptococcus, converts the pyruvic acid molecules into two molecules of lactic acid.

assimilated in the form of ammonium ions (NH_4^+) and by some species in the form of nitrate ions (NO_3^-). The ability of yeast to convert inorganic nitrogen into cellular proteins is being exploited to make microbial protein, sometimes referred to as single-cell protein, which can serve as a supplement in both human food and animal feed.

Let me briefly describe what yeasts are, how they reproduce and what their place is in the kingdom of the Mycota, or fungi. Because of yeasts' predominantly unicellular form of growth they are often described as microfungi. The more than 500 known species of yeasts are classified taxonomically according to their mode of reproduction, which can be sexual or asexual. Three classes of fungi accommodate the yeasts: the Ascomycetes, the Basidiomycetes and the Deuteromycetes.

The first class includes the yeast whose sexual reproductive structures take the form of simple asci and ascospores. A diploid yeast cell (a cell with two sets of paired chromosomes)

undergoes meiosis (reduction division) and forms four to eight ascospores, enclosed in an ascus, or sac. The ascospores are haploid cells (which have only one set of chromosomes); haploid cells of different sexes combine to form a new diploid organism. This mode of sexual reproduction can be manipulated by investigators to carry out genetic hybridizations for the improvement of strains. Numerous yeast species belong to the class Ascomycetes, but only a few are of industrial importance. By far the earliest yeast to serve man and the most intensively cultivated is *Saccharomyces cerevisiae,* specific strains of which are used for brewing, wine making, sake making, baking and the production of industrial alcohols. *Kluyveromyces fragilis* is a lactose-fermenting species exploited on a small scale for producing alcohol from whey. *Saccharomycopsis lipolytica,* the species capable of metabolizing hydrocarbons, is an industrial source of citric acid.

The class Basidiomycetes has supplied no yeasts of industrial importance. It includes certain mushrooms, toad-

stools, smuts and rusts. These organisms reproduce by forming external sexual spores on a structure termed a basidium or on a promycelium arising from a teliospore.

The class Deuteromycetes includes yeasts that have no known sexual mode of reproduction; they reproduce only vegetatively, which usually means by budding. The class has a few species of industrial significance. *Candida utilis,* the yeast that can grow on sulfite waste liquor, is one. Another, *Trichosporon cutaneum,* plays an important role in aerobic sewage-digestion systems because of its enormous capacity for oxidizing organic compounds, including some that are toxic to other fungi, such as phenolic compounds. A recently recognized species, *Phaffia rhodozyma,* has the ability to make an unusual carotenoid. Carotenoids are among the pigments found in plants. The carotenoid astaxanthin made by this yeast is being tested as a source of pigment for salmon and trout reared in pens. Astaxanthin gives the flesh of trout and salmon their normal orange pink color. The flesh of fish

CONVERSION OF METHANE INTO PROTEIN is carried out by the bacterium *Methylophilus methylotrophus.* Usually the methane (CH_4) is first oxidized with the aid of a catalyst to methanol (CH_3OH), which then serves as the substrate for bacterial growth. Enzymes oxidize the methanol to formaldehyde (H_2CO), which condenses with ribulose 5-phosphate to form 3-oxohexulose 6-phosphate. After a sequence of reactions the carbon and oxygen atoms in three molecules of methanol are transformed into one molecule of pyruvate (CH_3COCOO^-); ribulose 5-phosphate is released to renew cycle. Pyruvate is starting point for compounds needed for growth.

raised in captivity is white but will acquire a normal color if the fish are fed a source of the pigment.

Vegetative reproduction in the industrial species of yeast is mainly by the repeated formation of buds on the cell surface. A mother cell is said to give rise to a number of daughter cells, which bud in their turn. The shape of the vegetative cells varies from stubby oval to elongate. Most yeasts under industrial growing conditions propagate only vegetatively. The sexual cycle, if it exists at all, is induced only by special culture conditions.

Molds are filamentous fungi that form a large group of eukaryotic organisms lacking chlorophyll that together with the unicellular yeasts constitute the Mycota. Molds were recognized long before the yeasts because of their tendency to form a matlike somatic tissue, visible as spoilage organisms on many types of food. Everyone has seen moldy oranges and other fruits, moldy cereal and blue mold on cheese.

Sometimes the somatic tissue is not conspicuous but the sexual fruiting bodies are. For example, the vegetative growth of mushrooms is hidden in the soil and only the mushrooms or puffballs are visible aboveground. The case of bracket fungi on dying trees is similar. The vegetative growth of such fungi is gathering nutrients in the trunk of the tree from the cellulose or lignin components of the rotting wood and is therefore not visible. The spore-generating reproductive structures appear on the trunk and can weigh several pounds. Many diseases of both plants and animals are caused by fungi. Some fungi are obligate parasites, which means they depend on a living host to complete their life cycle, whereas others are opportunists. The latter normally live on dead organic matter but occasionally attack a living host, particularly one whose natural defenses are weakened for one reason or another.

In addition to having true nuclei and lacking the ability to conduct photosynthesis, the fungi (excluding the unicellular yeasts) are usually characterized as structures whose vegetative body or somatic tissue is filamentous and branched and typically has tough cell walls made up of variously linked polymers of glucose (known as glucans), of glucosamine (chitosan) and *N*-acetylglucosamine (chitin). In a few instances the cell wall consists entirely of chitin. Such a vegetative structure is known as a mycelium; the tubelike structures constituting the mycelium are called hyphae. The hyphae either can be separated into individual cells by septa or can be essentially free of septa, in which case the mycelium is termed coenocytic. Most fungal cells have many nuclei even when the hyphae consist of rows of individual cells divided by septa. Although some fungi fall outside the above definition, it covers the fungi of industrial interest.

Fungi can reproduce both sexually and asexually. The asexual fungi generate various kinds of unicellular asexual spores by division of the cell nucleus. The spores develop on sporophores, specialized structures that extend into the air from the vegetative mycelium. At the tips of the structures the spores themselves are borne. If they are enclosed in a sporangium, or saclike device, the spores are referred to as sporangiospores. Spores not enclosed in a sac are conidia. At maturity both sporangiospores and conidia are readily distributed by the wind. If they fall on a suitable substrate, they germinate and form new mycelia that can in turn generate new reproductive structures.

The morphology of the spore-bearing structures is highly variable and constitutes one of the bases on which fungi are classified. The somatic mycelium is usually not sufficiently distinctive to be of much help in classification. The color of most molds that live on decaying organic matter is due to the color of their asexual spores. They exhibit various shades of white, blue, green, red, brown or black.

Many fungi can also reproduce by forming sexual spores, generated by the meiosis, or reduction division, of a diploid nucleus. In meiosis the number of chromosomes is halved by the unpairing of homologous chromosomes. The sexual spores contain only one each of the pairs of homologous chromosomes. The diploid condition is reestablished when two haploid spores come together and fuse, thereby completing the life cycle. The fungi with no known sexual cycle are placed, like the yeasts, in the class Deuteromycetes; they are also known as fungi imperfecti.

Fungi of importance in industrial microbiology that are endowed with sexual reproductive structures fall into three classes: the Ascomycetes, the Basidiomycetes and the Zygomycetes. As with the yeasts, the ascomycetous fungi produce their spores in asci. In the filamentous true fungi, however, the asci are formed inside a complex fruiting body, the ascocarp. Similarly, the basidiomycetous fungi develop their sexual spores externally on basidia, which are enclosed in a complex fruiting body, the basidiocarp. Fungi belonging to the class Zygomycetes form sexual zygospores that are almost microscopic in size. Under natural conditions fungi reproduce for the most part asexually; sexual reproductive structures appear only occasionally under favorable circumstances. Fungi of industrial significance are grown mainly in tanks as submerged clumps of mycelium. Under such artificial conditions neither sexual nor asexual spores are generated.

The nutritional requirements of fungi closely follow those I have described for the yeasts except that the fungi (whose species outnumber those of the yeasts more than a hundredfold) are more diversified in the variety of organic substrates they can assimilate. For example, there are no yeasts that can grow on cellulose or lignin, whereas certain fungi can. On the other hand, many yeasts can conduct an active anaerobic fermentation of sugar into ethanol, whereas with few exceptions fungi are strict aerobes. The fungi will accept either organic or inorganic nitrogen, but neither fungi nor yeasts can assimilate nitrogen gas from the atmosphere, as some bacteria can. Fungi require a source of various minerals, particularly phosphate, sulfate and salts of potassium and magnesium. They also need a number of trace elements in salt form: boron, manganese, copper, molybdenum, iron and zinc. These elements are required for the proper functioning of various metabolic enzymes. Yeasts have similar requirements.

Fungi have great economic importance not only for their usefulness but also for the harm they do. Fungi are responsible for the destruction of much of the organic matter on the earth, a largely beneficial activity since it is integral to the recycling of living matter. On the other hand, fungi cause manifold diseases of plants and animals and can destroy foods and materials on which man depends. A small sampling will suggest the range of these effects. The Dutch elm disease is caused by an ascomycete, *Ceratocystis ulmi*. A water mold attacks fishes and fish eggs in hatcheries. *Coccidioides immites*, a deuteromycete, is responsible for coccidioidomycosis, or San Joaquin fever, in man and some animals. Cotton fabrics are destroyed by the cellulose-digesting ascomycete *Chaetomium*.

Fungi can also poison human food and animal feed. *Claviceps purpurea*, an ascomycete, elaborates a number of poisonous alkaloids when it parasitizes the rye plant, causing the disease known as ergotism. Consuming ergot-contaminated grain causes cattle to abort. In human beings ergotism can lead to hallucinations and death. Another form of poisoning is caused by fungi that secrete aflatoxins in improperly stored animal feed, such as peanut meal and hay; an example is the ascomycetous fungus *Aspergillus flavus*. Its toxins, which are secondary metabolites, are highly carcinogenic. The effects of poisonous mushrooms are well known.

The harmful effects of fungi are counterbalanced by their industrial uses. Fungi are the basis of many fermentations, such as the hydrolysis of rice starch that yields sake and the fermentations of various combinations of soybeans, rice and malt that yield the Oriental foods miso, shoyu and tempeh. The fungi are the source of many commercial enzymes (amylases, proteases, pectinases), organic acids (citric, lactic), antibiotics (notably penicillin), special cheeses (Camembert, Roquefort) and of course commercial mushrooms.

Let us return now to the simplest of all organisms, the prokaryotic ones, and consider their structure in somewhat more detail. The prokaryotes lack the organized nucleus, vacuoles, mitochondria and membrane systems present in yeasts, molds and other eukaryotes. The prokaryote cell has only two major internal features: a single closed loop of DNA and the nondifferentiated cytoplasm in which the DNA floats. The length of the DNA loop that encodes the cell's entire genetic blueprint is remarkable: it is more than a millimeter, or several hundred times the maximum dimension of most bacteria. The cytoplasm holds large numbers of ribosomes: granules made up of RNA and proteins that serve as the machines for assembling amino acids into proteins.

The ribosomes of prokaryotes are smaller than those of eukaryotes.

Prokaryotic microorganisms have cell envelopes very different from the envelopes of eukaryotic cells. All prokaryotic cell walls incorporate one common chemical component, peptidoglycan, which is responsible for most of the wall's shape and strength. Peptidoglycan is a large polymer built around alternating subunits of N-acetylglucosamine (which is also the building block of chitin in eukaryotes) and N-acetylmuramic acid. The latter molecule is similar to N-acetylglucosamine but has a unit of pyruvic acid attached to its third carbon atom. The pyruvic acid serves as an attachment point for a linear side chain consisting of the amino acids L-alanine, D-glutamic acid, diaminopimelic acid (an amino-acid-like compound) and D-alanine. These four-unit side chains serve to cross-link peptidoglycan molecules into a giant saclike molecule that surrounds the entire cell.

The peptidoglycan content of bacterial cell walls varies greatly from more than 50 percent in the wall of Gram-positive bacteria to less than 10 percent in Gram-negative bacteria. (The term Gram-positive refers to the ability of certain bacteria to hold the stain crystal violet, which Gram-negative bacteria do not retain.) The peptidoglycan content of cells is correlated with their sensitivity to penicillin because the antibiotic interferes with the biosynthesis of peptidoglycan. This explains why Gram-positive bacteria (which have a high peptidoglycan content) are much more sensitive to penicillin than Gram-negative bacteria. It also explains why only growing cells are killed by penicillin: in the presence of the antibiotic the cell wall under construction cannot be finished and the cell dies. Certain other antibiotics interfere with peptidoglycan biosynthesis in other ways.

In Gram-negative bacteria the inner layer of the cell wall is poor in peptidoglycan and the outer layer is rich in lipoproteins and lipopolysaccharides, which supply up to 80 percent of the wall's dry weight. The lipopolysaccharide component is mainly responsible for what is known as the O-antigenic specificity exhibited by Gram-negative bacteria. Each strain has a somewhat different lipopolysaccharide on its surface, depending on the particular sugars incorporated in the wall polymers. When human beings or other animals are infected or inoculated with such bacteria, the lipopolysaccharide elicits the formation of specific antibodies. With the aid of specific antiserums it is possible to identify the strain of bacteria that was responsible for a particular infection.

Many prokaryotic microorganisms are endowed with organelles that enable them to move about. The commonest structures are whiplike flagella that project from the cell surface. If the flagella are concentrated at one end of the cell, they are called polar; if they are distributed evenly over the cell surface, they are called peritrichous.

Prokaryotic cells have many shapes: spherical, rodlike and even (in the case of the actinomycetes) branched. Although some prokaryotic cells may equal or exceed the length of some eukaryotic microorganisms, the cellular volume of prokaryotes is always much less. Prokaryotic cells multiply asexually, almost always by the formation of a septum, or cross wall, after the chromosome has been replicated. The two ends of the cell, each with a chromosome, thereupon fission into two new cells. Some of the true bacteria (Eubacteria) contain endospores: packages of DNA that can remain dormant for many years and are highly resistant to heat (even boiling water), toxic chemicals and other insults that kill the vegetative cell. Then, under the appropriate conditions, the endospore can give rise to a new bacterium.

Although a small number of pro-

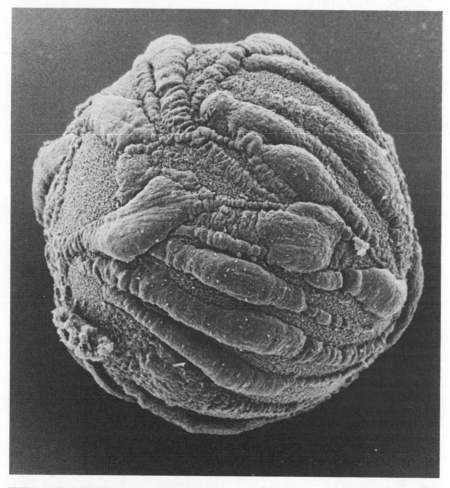

HUMAN FORESKIN CELLS used to produce interferons in tissue culture are the elongated protuberances on the spherical surface in this scanning electron micrograph. The sphere is a tiny bead of the synthetic polymer dextran; such beads are introduced into the culture to provide a surface on which the cells can grow. The micrograph, which enlarges the sphere 1,100 diameters, was made by Don Siegel of Harvard University and Robert Fleischaker of M.I.T.

karyotic microorganisms can conduct photosynthesis with the aid of a special kind of bacterial chlorophyll, most prokaryotes, including those of industrial importance, are heterotrophic: they require a source of assimilable carbon and nitrogen, together with various mineral salts and in some instances organic growth factors, for example vitamins. The prokaryotes of industrial utility grow on an organic substrate that serves as a source of both carbon and energy. They can usually synthesize all their cell constituents from a single organic compound and a source of nitrogen. Some bacteria can "fix" the gaseous nitrogen of the atmosphere, that is, convert it into organic nitrogen. Examples are species of *Azotobacter*, which are free-living in the soil, and *Rhizobium*, which grow symbiotically in the root nodules of leguminous plants. A major goal of agricultural technologists is to find a way to incorporate the nitrogen-fixing genes of such bacteria into the cells of corn and other plants whose nitrogen needs must now be met with fertilizers.

In general the metabolism of prokaryotic microorganisms is strongly influenced by molecular oxygen. Organisms that depend on aerobic respiration and for which oxygen functions as the terminal oxidizing agent are classed as obligate aerobes. The antibiotic-secreting *Streptomyces* are an example. In contrast, for organisms that are obligate anaerobes oxygen is usually toxic. The intermediate group of organisms, the facultative anaerobes, can grow in either the presence or the absence of molecular oxygen. Such organisms can be subdivided into two groups, depending on whether oxygen is actively metabolized or is merely tolerated. The lactic acid bacteria belong to the group that get their energy entirely from fermentation, yet these bacteria are not harmed by oxygen. On the other hand, the coliform bacteria, such as the bacterium *Escherichia coli*, can obtain their energy either from fermentation or respiration.

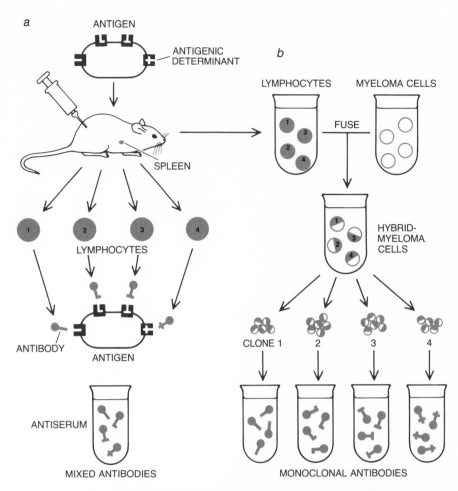

MONOCLONAL ANTIBODIES are ultrapure antibodies manufactured by hybridoma cells: fused lymphocytes (white blood cells) and malignant myeloma cells. At the top left is a schematic representation of an antigen with four antigenic determinants on its surface. When the antigen is injected into a mouse, lymphocytes of the mouse separately manufacture antibodies specific for the antigenic determinants. Therefore an antiserum prepared from the blood of the mouse contains a mixture of antibodies against the antigen. In the preparation of monoclonal antibodies, shown at the right, lymphocytes are removed from the spleen of the mouse and allowed to fuse with myeloma cells. Hybrid cells can be cloned to produce pure antibodies.

Among the bacteria only one subgroup, the Eubacteria, provides species of industrial usefulness. These true bacteria constitute such a large and diversified group of organisms that a complete taxonomic treatment would be out of place here. I shall therefore give only a few examples of the Eubacteria exploited by industry. The acetic acid bacteria, represented by the genera *Gluconobacter* and *Acetobacter*, are Gram-negative, rod-shaped organisms that can convert ethanol into acetic acid. *Gluconobacter* has polar flagella and oxidizes ethanol only as far as acetic acid. *Acetobacter* has peritrichous flagella and is capable of oxidizing the acetic acid it forms into carbon dioxide and water.

Aerobic spore-forming bacteria are represented by the genus *Bacillus*, which has found significant use in industrial fermentations. All species of bacilli are able to form endospores and are Gram-positive when the cells are young. Nearly all the species have peritrichous flagella. Some species, such as *Bacillus subtilis*, are strict aerobes; others, such as *B. thuringiensis*, can also conduct anaerobic fermentation. *B. subtilis* has certain attributes that make it attractive as a replacement organism for *E. coli*, which is now almost the universal organism of recombinant-DNA fermentations. It appears that secondary metabolites elaborated by *B. subtilis* are secreted by the cell and therefore readily collected whereas those formed by *E. coli* remain inside the cell and can be obtained only by breaking down the cell and isolating the desired product from the debris. *E. coli* also has the disadvantage of containing highly toxic substances (endotoxins) that must be carefully removed from the product of the fermentation. *B. subtilis* does not manufacture such endotoxins.

Anaerobic spore-forming bacteria are represented by species of the genus *Clostridium*. Whereas vegetative cells of the species are highly sensitive to oxygen, the endospores are protected from this lethal effect. The cells, common in soil, are Gram-positive rods with peritrichous flagella. They obtain their metabolic energy from various types of fermentation. I have mentioned that *C. acetobutylicum* can ferment sugars into acetone, ethanol, isopropanol and butanol; other fermentable substrates for this group include starch, pectin and various nitrogenous compounds.

The lactic acid bacteria include, among others, species of the genera *Streptococcus*, *Leuconostoc* and *Lactobacillus*. These organisms, which do not form endospores, are Gram-positive, nonmotile rods or spheres that get their energy from fermentation but are not sensitive to oxygen. The heterofermentative lactic acid bacteria of the genus *Leuconostoc* convert carbohydrates into lactic acid, ethanol and carbon dioxide. The homofermentative lactic acid bacteria of the genus *Streptococcus* yield only lactic acid. Species of *Lactobacillus*

ORGANISM	TYPE	PRODUCT
FOODS AND BEVERAGES		
Saccharomyces cerevisiae	YEAST	BAKER'S YEAST, WINE, ALE, SAKE
Saccharomyces carlsbergensis	YEAST	LAGER BEER
Saccharomyces rouxii	YEAST	SOY SAUCE
Candida milleri	YEAST	SOUR FRENCH BREAD
Lactobacillus sanfrancisco	BACTERIUM	SOUR FRENCH BREAD
Streptococcus thermophilus	BACTERIUM	YOGURT
Lactobacillus bulgaricus	BACTERIUM	YOGURT
Propionibacterium shermanii	BACTERIUM	SWISS CHEESE
Gluconobacter suboxidans	BACTERIUM	VINEGAR
Penicillium roquefortii	MOLD	BLUE-VEINED CHEESES
Penicillium camembertii	MOLD	CAMEMBERT AND BRIE CHEESES
Aspergillus oryzae	MOLD	SAKE (RICE-STARCH HYDROLYSIS)
Rhizopus	MOLD	TEMPEH
Mucor	MOLD	SUFU (SOYBEAN CURD)
Monascus purpurea	MOLD	ANG-KAK (RED RICE)
INDUSTRIAL CHEMICALS		
Saccharomyces cerevisiae	YEAST	ETHANOL (FROM GLUCOSE)
Kluyveromyces fragilis	YEAST	ETHANOL (FROM LACTOSE)
Clostridium acetobutylicum	BACTERIUM	ACETONE AND BUTANOL
Aspergillus niger	MOLD	CITRIC ACID
Xanthomonas campestris	BACTERIUM	POLYSACCHARIDES
AMINO ACIDS AND FLAVOR-ENHANCING NUCLEOTIDES		
Corynebacterium glutamicum	BACTERIUM	L-LYSINE
Corynebacterium glutamicum	BACTERIUM	5'-INOSINIC ACID AND 5'-GUANYLIC ACID
SINGLE-CELL PROTEINS		
Candida utilis	YEAST	MICROBIAL PROTEIN FROM PAPER-PULP WASTE
Saccharomycopsis lipolytica	YEAST	MICROBIAL PROTEIN FROM PETROLEUM ALKANES
Methylophilus methylotrophus	BACTERIUM	MICROBIAL PROTEIN FROM GROWTH ON METHANE OR METHANOL
VITAMINS		
Eremothecium ashbyi	YEAST	RIBOFLAVIN
Pseudomonas denitrificans	BACTERIUM	VITAMIN B_{12}
Propionibacterium	BACTERIUM	VITAMIN B_{12}
ENZYMES		
Aspergillus oryzae	MOLD	AMYLASES
Aspergillus niger	MOLD	GLUCAMYLASE
Trichoderma reesii	MOLD	CELLULASE
Saccharomyces cerevisiae	YEAST	INVERTASE
Kluyveromyces fragilis	YEAST	LACTASE
Saccharomycopsis lipolytica	YEAST	LIPASE
Aspergillus	MOLD	PECTINASES AND PROTEASES
Bacillus	BACTERIUM	PROTEASES
Endothia parasitica	MOLD	MICROBIAL RENNET
POLYSACCHARIDES		
Leuconostoc mesenteroides	BACTERIUM	DEXTRAN
Xanthomonas campestris	BACTERIUM	XANTHAN GUM
PHARMACEUTICALS		
Penicillum chrysogenum	MOLD	PENICILLINS
Cephalosporium acremonium	MOLD	CEPHALOSPORINS
Streptomyces	BACTERIUM	AMPHOTERICIN B, KANAMYCINS, NEOMYCINS, STREPTOMYCIN, TETRACYCLINES AND OTHERS
Bacillus brevis	BACTERIUM	GRAMICIDIN S
Bacillus subtilis	BACTERIUM	BACITRACIN
Bacillus polymyxa	BACTERIUM	POLYMYXIN B
Rhizopus nigricans	MOLD	STEROID TRANSFORMATION
Arthrobacter simplex	BACTERIUM	STEROID TRANSFORMATION
Mycobacterium	BACTERIUM	STEROID TRANSFORMATION
HYBRIDOMAS	——	IMMUNOGLOBULINS AND MONOCLONAL ANTIBODIES
MAMMALIAN CELL LINES	——	INTERFERON
Escherichia coli (via recombinant-DNA technology)	BACTERIUM	INSULIN, HUMAN GROWTH HORMONE, SOMATOSTATIN, INTERFERON
CAROTENOIDS		
Blakeslea trispora	MOLD	BETA-CAROTENE
Phaffia rhodozyma	YEAST	ASTAXANTHIN
ENTOMOPATHOGENIC BACTERIA		
Bacillus thuringiensis	BACTERIUM	BIOINSECTICIDES
Bacillus popilliae	BACTERIUM	BIOINSECTICIDES

ferment sugars into various products along with lactic acid.

Another prokaryote, *Corynebacterium glutamicum,* as I have noted, is a major industrial source of lysine and of flavor-enhancing 5'-nucleotides. The coryneform bacteria tend to be irregular in shape; they are sometimes branched rather than simply club-shaped (as the Greek *coryne,* club, suggests). The cells are generally nonmotile, Gram-positive and lack the ability to form endospores. The genus contains species that are pathogenic to animals and plants, but there are also soil-inhabiting species that are nonpathogenic and of industrial interest. Although the cells are facultative anaerobes, they grow best aerobically. Corynebacteria make a catalase enzyme that decomposes hydrogen peroxide (H_2O_2) into water and oxygen.

Another large group of prokaryotic soil organisms, the true actinomycetes, are strict aerobes with simple nutritional requirements. They include many genera whose vegetative development is exclusively mycelium. They are Gram-positive and do not form endospores. By far the largest genus in the group is *Streptomyces,* whose species assumed major importance when it was discovered that they secrete useful antibiotics. The characteristic smell of damp forest soil is caused by volatile compounds elaborated by *Streptomyces.* When actinomycetes are grown on a solid medium, they not only form a finely branched mycelium but also give rise to aerial hyphae that differentiate into chains of conidiospores. Each conidiospore in turn can generate a mycelial microcolony. Another genus of the true actinomycetes is *Micromonospora,* some of whose species also secrete antibiotics. Their colonies are devoid of aerial mycelia. Instead conidiospores are formed singly at the tips of short hyphal branches throughout the colony.

Human cells were first cultured in laboratory glassware early in this century. No significant industrial use was found for mammalian tissue cultures, however, until the early 1950's; then it was found that the virus of poliomyelitis could be grown in cultures of monkey and human tissue for the manufacture of vaccines. Interest in human cell lines has greatly increased since then with the application of cell cultures in the isolation and growth of other viruses, in the production of highly specif-

MAIN INDUSTRIAL PRODUCTS obtained with the help of microorganisms range from some of the oldest (beer, cheese and leavened bread) to the newest creations of recombinant-DNA technology (insulin, human growth hormone). Cell technology has recently been broadened to include the culture of mammalian cells as sources of new products.

ic proteins (such as interferon and antibodies), in cancer research and in antiviral chemotherapy.

Mammalian cells are of course eukaryotic and are generally more complex in their internal organization than fungal or yeast cells. One fundamental difference between mammalian cells and the cells of microorganisms is that mammalian cells have no tough outer wall; their cytoplasm is enclosed only by a thin membrane. This plasma membrane regulates the uptake of nutrients needed for cell maintenance and division and the release of cellular metabolic products. In somatic tissues (tissues other than reproductive ones) the cells are diploid and divide by mitosis. Constriction of the membranes serves to divide the cell in two. Cellular division takes on the order of 24 hours compared with one and a half to two and a half hours for yeasts and 20 to 60 minutes for bacteria. Normally mammalian cells are arranged in three-dimensional structures such as organs and muscles, but when they are grown in tissue culture, they can float free or form a layer one cell thick on a surface.

The nutritional requirements of mammalian cells are a good deal more complex than those of eukaryotic microorganisms. Mammalian cells must be supplied with a mixture of amino acids for the synthesis of proteins and with purines and pyrimidines for the synthesis of nucleic acids. The growth medium, which must be made up with very pure deionized water, must contain glucose as a source of carbon and energy, a mixture of vitamins and a balanced mixture of minerals to maintain the cells at the appropriate osmotic pressure and to buffer the medium at the optimum pH (about 7.2). The medium must also be supplied with small amounts of antibiotics to control bacterial infection and must consist of 5 to 20 percent blood serum (either human or fetal bovine). In some cases specific serum proteins can serve as a satisfactory replacement for whole serum when it is necessary to have a culture free of serum. For optimum growth the culture must be held close to 37 degrees Celsius. Below 36 degrees C. the cells divide very slowly or not at all; above 38 degrees they die. Most mammalian cell lines, including human ones, can be stored indefinitely if they are cooled slowly to −180 degrees. Special media are needed for frozen storage.

Mammalian cell lines are commonly started from embryonic tissue. The usual procedure is to obtain a suspension of single cells by treating the dispersed tissue with the digestive enzyme trypsin. If a suspension of tissue cells in a nutrient medium is allowed to settle onto the flat surface of a culture vessel, the cells flatten out and divide to form a layer one cell thick. When growth is well established, subcultures can be started by

carefully breaking up clumps or sheets of cells. Some types of cells can also be grown in suspension. The usual technique employs cylindrical roller bottles that are rotated slowly about their long axis. Cell yields can be enhanced by adding to the suspension small microcarrier beads made from inert synthetic polymers. Submerged cultures have also been achieved in agitated vessels with a capacity of more than 1,000 liters.

A small number of therapeutically useful proteins and polypeptides, such as human insulin and somatostatin (a peptide chain consisting of only 14 amino acid units), are being obtained from bacterial cells, notably *E. coli,* into which the gene for making the desired product has been introduced by recombinant-DNA techniques. The separation of such proteins from microbial proteins and other cell constituents, many of them highly toxic, presents major difficulties. The removal of foreign proteins can be avoided or simplified by the use of mammalian (including human) cell lines that have been modified to increase their productivity.

One protein that is currently being made on a substantial scale (although it is still a laboratory one) from mammalian cell lines is the antiviral agent interferon. There are many kinds of interferon, not only interferons from different mammalian species but also different interferons from the same species. One of the chief problems that had to be overcome in obtaining interferon from cultured cells is the extremely low concentration of interferon made by cells. The production process starts with the growing of a particular cell line for about a week in a nutrient medium. At that stage little or no interferon is being synthesized. The nutrient is then replaced with an inducer medium that typically contains a mixture of polyinosine-cytosine RNA (a synthetic double-strand RNA) and diethylaminoethyl dextran. The two compounds induce the cells to start manufacturing interferon. Before the cells begin to excrete interferon the medium is replaced once more with a medium containing additional substances (such as insulin and guanosine phosphate or low concentrations of serum albumin) that have been found to increase the yield of interferon or to increase its stability. Finally the medium is collected, concentrated, dialyzed and freeze-dried.

The resulting material contains only about .1 percent of pure interferon, so that other purification methods must come into play. One of the most effective and specific is immunoaffinity chromatography. Monoclonal antibodies with an affinity for a particular type of interferon can be attached to polysaccharide beads, which are placed in a glass column. When the crude interferon solution is passed through the col-

umn, the interferon molecules are adsorbed on the beads while the impurities pass through the column. The interferon is released from the beads and eluted from the column by altering the pH of the column with a suitable washing solution. In a single passage through the column the interferon activity can be raised some 5,000-fold.

Monoclonal antibodies have themselves only recently become available in industrial quantities through techniques in which normal mammalian cells are hybridized with myeloma cells of malignant tumors of the immune system. The normal immune system is capable of manufacturing at least a million different kinds of antibodies to combat and inactivate foreign proteins or other antigens that may invade the body. A malignant myeloma cell, however, synthesizes only a single type of antibody, an immunoglobulin protein that may be any one of the almost innumerable proteins possible. Myeloma cells proliferate rapidly and can be cultured indefinitely from a single cell. They cannot, however, be induced to yield antibodies to a specific antigen.

That difficulty was overcome in 1975 when Cesar Milstein and his colleagues at the Medical Research Council Laboratory of Molecular Biology in Cambridge, England, conceived the idea of fusing mouse myeloma cells with *B* lymphocytes from the spleen of a mouse immunized with a specific antigen. The resulting "hybridoma," or hybrid myeloma, cells had the properties of both parent cells: immortality and the ability to secrete large amounts of a single, specific type of antibody. Many details in the selection of hybrid cells had to be worked out, including the genetic properties of the hybrid animal cells.

Milstein's work has opened a new era of experimental immunology. The problems previously associated with heteroantiserums, that is, serums containing mixtures of antibodies, could in principle now be circumvented. In 1980 Carlo M. Croce and his co-workers at the Wistar Institute of Anatomy and Biology in Philadelphia succeeded in generating a stable, antigen-producing intraspecific human hybridoma by fusing *B* lymphocytes from a patient suffering from multiple myeloma with peripheral lymphocytes from a patient with subacute panencephalitis. The hybridoma cells of this fusion were found to secrete molecules of human immunoglobulin *M* specific for components of the measles virus. Although only a limited number of human cell lines lead to hybridomas that actively secrete specific antibodies, Croce's work indicates the possibility of obtaining human *B*-cell hybrids that continuously secrete human antibodies against a variety of pathogenic viruses. The potential for improving human immunotherapy is therefore great.

3

The Genetic Programming of Industrial Microorganisms

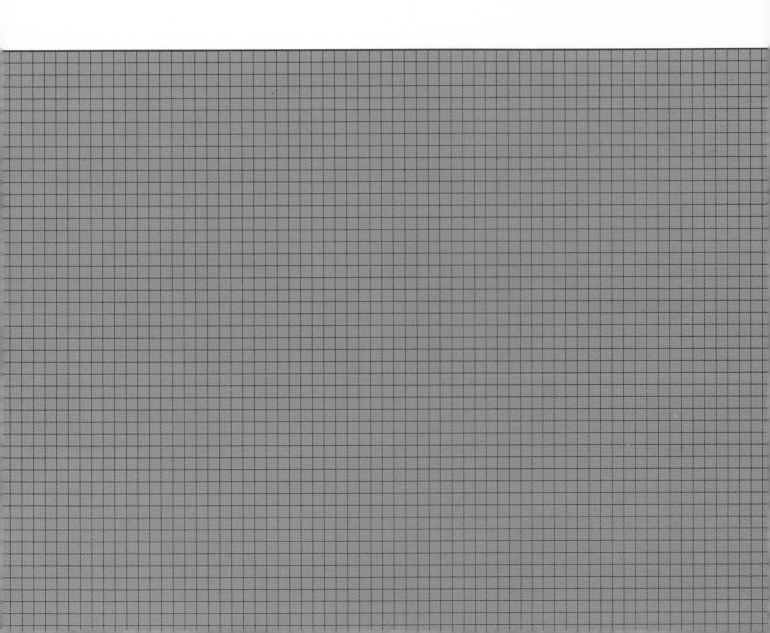

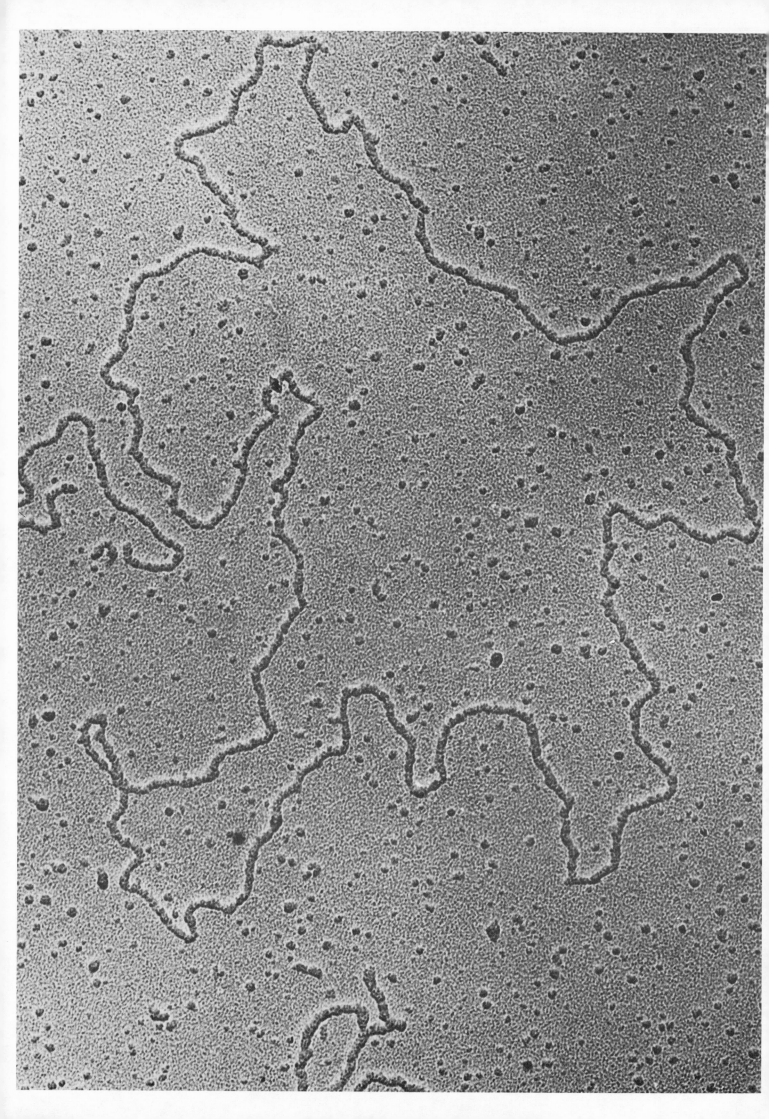

The Genetic Programming of Industrial Microorganisms

BY DAVID A. HOPWOOD

The useful products made by microorganisms are specified by genes. The genes in turn are specified by intensive selection, and now by direct intervention such as introducing genes from other organisms

A microorganism is a finely integrated machine that has evolved to serve its own purposes: survival and reproduction. A "wild type" bacterium or yeast cell has become closely adapted through natural selection to its environment and to competition with other species; it is not adapted to the manufacture of some substance that happens to be sought after by man. Modern industrial microbiology calls for the selection or construction of freak organisms, genetically programmed to make a normal metabolic product in amounts that would be a disastrous drain on a wild organism's resources of energy and nutrients or even to make a product that is not part of its normal repertory.

The first steps toward controlling and improving microbiological processes were taken only a little more than 100 years ago, when bacteria and fungi that made desired commodities were isolated and grown in pure cultures and it became possible to select strains particularly suited to a given task.

The purposeful breeding of special industrial strains became possible only later, as something was learned about microbial genetics. First came the discovery of some of the mechanisms of mutation: the sudden change of a gene, the unit of hereditary information, into a new form. Mutations were induced in the laboratory by means of X rays as early as 1927, and the discovery after 1945 of a wide range of other potent mutagenic radiations and chemical mu-

tagens gave microbiologists a powerful set of tools for changing the genetic composition of their cultures. The mid-1940's also saw advances in genetics that made it possible to reshuffle genetic information by recombining genes from two or more organisms: bacteria were found to reproduce sexually by a bizarre form of mating and even by the exchange of naked DNA, and novel genetic systems were discovered in fungi. Improved understanding of these processes initiated the explosive advance in microbial genetics and molecular biology that is still under way today.

In the years after World War II the fermentation industry underwent important changes, both in its capabilities and in the volume of its output, with the industrial production of antibiotics. Penicillin had been manufactured during the war, and it was followed by a growing list of new antibiotics effective against a broad range of bacterial and fungal diseases. Then new fermentations were developed in which microorganisms yielded other pure chemicals such as amino acids (the components of proteins) and nucleotides (the components of DNA). Such chemicals could not be manufactured economically by wild-type organisms. Their industrial production depended on genetic manipulation, and so the new fermentation industry and a new science of microbial genetics developed in parallel. For a long time, however (and to the frustration of some academic geneticists), it was the exception rather than the rule

for the science of genetics to make an appreciable contribution to the genetic programming of industrial microorganisms.

That situation changed dramatically after the announcement in 1973 of experiments involving recombinant DNA and molecular cloning. New techniques were developed that make it possible (in principle) to transfer genes from any source into any microorganism. These techniques of genetic engineering are powerful laboratory tools for revealing the structure and function of genes. And they have immense potential for the breeding of industrial strains of microorganisms that can make such completely new fermentation products as human insulin or growth hormone and also for the rational development of new strains better suited to making traditional fermentation products. Genetic engineering has caught the imagination of industrial managers and entrepreneurs, but it is only the capstone of an edifice of microbial genetics built over the past 35 years; it is only one (albeit the most exciting) of the many facets of the genetic programming of industrial microorganisms.

Genetic information is stored in living cells in the threadlike molecule of DNA. In a typical bacterium the basic set of information is encoded in a single long, tangled molecule of DNA: the bacterial chromosome. It is a double helix each strand of which is a chain of nucleotides. Each nucleotide is characterized by one of four bases: adenine (A), guanine (G), thymine (T) and cytosine (C). The bases are complementary: an adenine in one strand pairs with a thymine in the other strand, and guanine pairs with cytosine. The molecule is a closed loop at least a millimeter in circumference (tightly folded to fit inside a bacterial cell perhaps a thousandth of a millimeter in diameter) consisting of several billion base pairs. The total information content of the bacterium is

PLASMID isolated from a bacterium is enlarged 115,000 diameters in the electron micrograph on the opposite page. Plasmids are small circular (closed loop) molecules of the genetic material DNA that exist outside the bacterial chromosome. They play a major role in the genetic programming of microorganisms because they can be transferred from one strain to another (even of a different species) and can serve as vectors for introducing completely new genetic information, by means of recombinant-DNA techniques, into bacteria. This plasmid was isolated from *Streptomyces coelicolor* by Mervyn Bibb in the author's laboratory and was shadowed with platinum, exaggerating its thickness; the double helix of DNA is about 10 micrometers in circumference, comprising about 30,000 base pairs of DNA, enough for some 30 genes.

in a set of several thousand "structural genes" (each one perhaps 1,000 base pairs long) that specify the same number of proteins, mostly enzymes. The information is encoded in the sequence of the bases on one strand of DNA, each three-base "codon" specifying one of the 20 amino acids; a structural gene directs the cell's machinery to assemble some hundreds of amino acid units into a particular linear sequence to form a particular protein.

Not all of the DNA has this coding function. Base sequences adjacent to the structural genes control their expression: their transcription into a messenger RNA complementary to the DNA template and the translation of the messenger RNA into protein on the cellular organelles called ribosomes. Two control regions regulate transcription. One is the promoter, a short sequence that enables the enzyme RNA polymerase to bind to the DNA and move along it, initiating transcription of the coding strand into RNA at a point before the beginning of the structural gene; the other control region is a signal to terminate transcription at the end of the structural gene. In genes that respond to the current concentration of particular metabolites in the cell (or in a culture medium) additional sites in the promoter and terminator regions interact with DNA-binding regulatory proteins. For example, an "operator" sequence between the promoter and the structural gene may bind a "repressor" protein, itself the product of a specific regulator gene; the

binding of a repressor, perhaps only in the presence of a particular metabolite, prevents the transcription of the structural gene. Other sequences, having been transcribed into messenger RNA, control translation. A "ribosomal binding site" fixes the RNA to the ribosome, allowing translation to begin at a "start" signal, the first codon of the structural gene; a "stop" signal at the end of the gene triggers the release of the completed protein chain.

Rational genetic programming depends on a clear understanding of these characteristics of DNA and of their differences in various organisms. Changes within a coding region, for example, can alter the amino acid sequence of an enzyme and thereby affect its activity. The slight alteration of a promoter sequence can increase the probability that RNA polymerase will bind to the promoter, and so enhance the rate of transcription. Mutations in operator regions or in a regulator gene can prevent the binding of a repressor and thereby greatly increase (derepress) transcription. Moreover, genes transplanted from one organism to an unrelated organism will be expressed only if the promoters and ribosomal-binding sites of the two organisms are similar enough.

The genetic code and the essential biochemistry of transcription and translation are the same in prokaryotes (bacteria) and eukaryotes (all higher organisms, from algae to man), but the control signals are different. So is the organization and expression of the DNA. In eu-

karyotes the DNA is complexed with protein and divided among a number of discrete chromosomes grouped within a membrane-bounded nucleus. Such eukaryotic organisms as fungi have perhaps 10 times as much DNA as bacteria; the higher plants and animals have thousands of times as much. The increase in DNA content is far in excess of the increase in the number of genes, in part because many genes in eukaryotes are split: noncoding intervening sequences ("introns") lie within the structural genes. Introns are transcribed along with the coding sequences ("exons") but are not expressed; they are removed in a splicing process that brings a gene's exons together to form the mature messenger RNA that is translated into protein. Introns are rare in the genes of fungi, but they are present in most genes of the higher eukaryotes. Their presence complicates recombinant-DNA manipulations because a bacterium lacks the enzymes for splicing them out of the primary RNA transcript, so that a natural eukaryotic gene containing introns cannot be expressed in bacteria.

The development of a tailor-made industrial microorganism from a wild bacterium or fungus calls for changing its genetic information in a way that eliminates undesirable properties, accentuates desirable properties or introduces entirely new ones. There are several ways to bring such changes about. One way is to take advantage of mutations or to induce them. The simplest

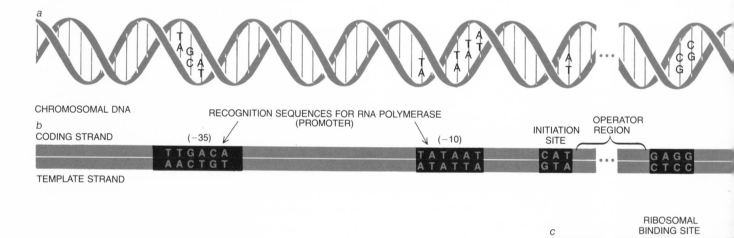

GENETIC INFORMATION is stored in the double helix of DNA (*a*). Each strand of the helix is a chain of nucleotides, each comprising a deoxyribose sugar and a phosphate group, which form the strand's backbone, as well as one of four bases: adenine (*A*), guanine (*G*), thymine (*T*) and cytosine (*C*). The information is encoded in the sequence of the bases along a strand. The complementarity of the bases (*A* always pairs with *T*, and *G* with *C*) is the basis of the replication of DNA from generation to generation and of its expression (shown here for bacterial DNA) as protein. Expression begins with the transcription of the DNA base sequence (*b*) into a strand of messenger RNA (*c*), which corresponds to the coding strand of the DNA except for the fact that uracil (*U*) replaces thymine. Transcription into RNA and translation into protein are regulated by special sequences (*black*) in the DNA and RNA respectively. The transcribing enzyme, RNA

kind of mutation is a point mutation: the change of one base pair (say adenine-thymine) to another (guanine-cytosine). In other instances a base pair or a short stretch of DNA may be deleted from a sequence, or a new base pair may be inserted. These changes are natural occurrences in any DNA, probably as the result of errors in its replication from generation to generation, but spontaneous mutations are rare, affecting a given base pair only about once in 100 million replications. The frequency of mutation can be increased at least a thousandfold by exposing microorganisms to such mutagens as ultraviolet radiation, ionizing radiation (X rays, gamma rays or neutrons) and a host of apparently unrelated chemical compounds that can react with DNA bases or interfere with DNA replication.

Mutagens hit genes at random. Each agent may act specifically on particular bases or groups of bases (for example, ultraviolet radiation tends to link two adjacent thymines on the same strand of DNA), but all genes are chains of the same bases. It is therefore usually impossible to cause a particular gene to mutate preferentially (even though in any gene particular treatments tend to cause mutations primarily at certain positions). To improve a strain by mutation one must rely on sensitive tests that make it possible to recognize the rare mutants that happen to have a desired characteristic.

Some procedures are straightforward. To select mutants resistant to a chemical that inhibits the growth of unmutated strains, for example, one can spread many millions of cells of the starting strain on a culture plate containing the inhibitor; only the resistant mutants proliferate and form colonies. Other kinds of mutants can be found only by testing the properties of random colonies in individual culture vessels or even in small versions of industrial fermenters; in such instances potent mutagens are required so that one can hope to find the desired mutant by examining thousands, rather than millions, of individuals. Two very different strategies for improving industrial microorganisms through mutational reprogramming are illustrated by examples involving the production of amino acids and antibiotics.

Lysine is an essential amino acid in animal nutrition (one that the animal cannot synthesize itself), but many plant proteins are deficient in it. Lysine is therefore produced by fermentation to serve as a supplement to animal feeds. The fermentation is based on an understanding in detail of both the pathway leading to bacterial biosynthesis of the amino acid and the pathway's genetic regulation. Skillful exploitation of that knowledge has made it possible to select mutant strains of *Brevibacterium flavum* and *Corynebacterium glutamicum* that convert more than a third of the sugar in a fermentation medium into lysine, yielding concentrations of as much as 75 grams of lysine per liter of medium. In these bacteria lysine is one end product of a branched pathway that also leads to the synthesis of the amino acids methionine and threonine. The main control ensuring the synthesis of enough of these amino acids to meet the needs of the bacterium, but not too much, is a feedback inhibition of the first enzyme of the pathway, aspartate kinase, by threonine and lysine acting together. That is, the accumulation of these two amino acids in excess of the organism's requirements tends to shut down their synthesis by inhibiting the enzyme's activity; conversely, a shortage of threonine or lysine increases their rate of synthesis.

Overproduction of lysine, far beyond the bacterium's own needs, has been achieved by isolating two kinds of mutant. In one kind a mutation in the gene coding for the enzyme homoserine dehydrogenase abolishes the enzyme's activity and thereby prevents the bacterium from making threonine (one of the inhibitory products) and methionine. When this auxotroph, or nutritionally deficient mutant, is cultured in a medium with just enough threonine and methionine to support growth but not enough threonine to cooperate with lysine in shutting off aspartate kinase's activity, the pathway to lysine continues to operate at full speed. Auxotrophic mutants are selected by testing thousands of colonies that have been treated with a mutagen; auxotrophs grow only if particular growth factors (in this case methionine and threonine) are supplied in the medium. Auxotrophs can be iso-

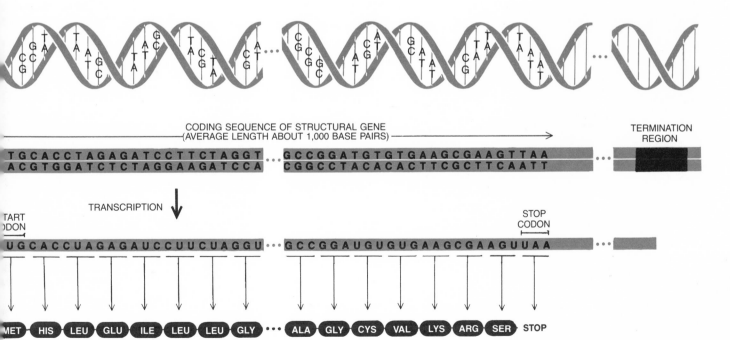

polymerase, binds to a promoter region that (in the bacterium *Escherichia coli*) has the specific sequences shown (or minor variations of them) about 10 base pairs and 35 base pairs before a transcription-initiation site; beyond the end of the structural gene a termination region causes the polymerase to cease transcription. In some genes an operator sequence, which can bind a repressor molecule, provides an extra control. Messenger RNA is translated on the cellular organelles called ribosomes; each triplet of bases (codon) encodes a particular amino acid and specifies its incorporation into the growing protein chain. A ribosomal binding site on the RNA allows translation to begin at a "start" codon, which is always *AUG* for the amino acid methionine (*Met*). Translation proceeds until a "stop" codon is reached (*UAA* is one of three possibilities) that signals the end of translation and detachment of completed protein chain from ribosome.

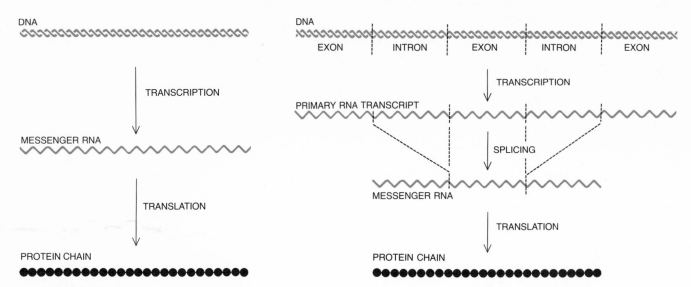

EXPRESSION OF DNA is different in prokaryotes (bacteria) and eukaryotes (higher organisms). In prokaryotes the genetic information, encoded in a continuous stretch of the DNA double helix that constitutes a structural gene, is transcribed directly into messenger RNA, which is translated to make a protein. In eukaryotes, on the other hand, some structural genes (most of them, in the higher eukaryotes) are split: coding sequences ("exons") are separated by noncoding intervening sequences ("introns"). Entire gene is transcribed to make primary RNA transcript. Then intron transcripts are excised and exon transcripts are spliced together to make messenger RNA.

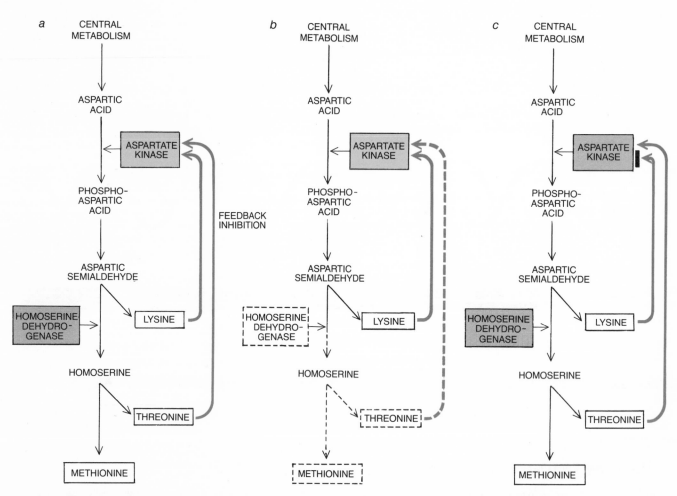

MUTATIONAL REPROGRAMMING yields bacterial strains that produce large amounts of the essential amino acid lysine. In wild-type strains (*a*) lysine is one product, along with the amino acids threonine and methionine, of a branched pathway controlled primarily by feedback inhibition: the activity of the enzyme aspartate kinase is inhibited by excess quantities of lysine and threonine acting together. The control is circumvented in two kinds of mutant strains. In one kind (*b*) a mutation inactivates the enzyme homoserine dehydrogenase, thereby preventing threonine from accumulating and inhibiting aspartate kinase. In the other kind (*c*) the gene coding for aspartate kinase itself is mutated; the altered enzyme functions, but it is not inhibited by lysine even when lysine is present in excess.

lated more efficiently by inoculating a mutagenized culture into a medium that lacks the appropriate growth factors and contains penicillin, which kills only proliferating bacteria; the auxotrophs cannot grow and so they survive, whereas the unmutated bacteria are killed.

A different kind of mutant has an altered form of aspartate kinase itself, which performs well enough as an enzyme but does not react with lysine and has therefore lost its sensitivity to feedback inhibition; again, high levels of lysine accumulate in the fermenter. These mutants could be selected because they are resistant to a compound called AEC, which resembles lysine so closely that it mimics its regulatory effect, inhibiting aspartate kinase even if no lysine is being synthesized. In the presence of AEC, wild strains die as a result of lysine starvation, whereas the mutants proliferate and form colonies. Lysine production is just one example of an industrial process that depends on the rational selection of mutants in which the precise controls regulating amino acid production are disconnected.

Antibiotics are very different. They are synthesized only at particular stages in the life cycle of certain molds, actinomycetes (filamentous bacteria) and spore-forming bacteria. Their genetic regulation is as yet poorly understood, but the amount of antibiotic produced certainly depends on many factors. Control systems respond to various aspects of the cell's metabolism (the availability of carbon, nitrogen and free phosphate, for example), to the synthesis of chemical building blocks needed to make the antibiotic and to the organism's resistance to its own potentially toxic antibiotic. Because the yield of an antibiotic depends on hundreds of genes, it is impossible to find individual mutations that can raise the yield from a wild strain's few milligrams per liter to an economic level, such as the 20 grams or more per liter of penicillin or tetracycline now being recovered from highly developed industrial strains respectively of *Penicillium chrysogenum* or *Streptomyces aureofaciens.*

These truly weird strains have been developed through many successive rounds of mutation and selection. In each round a culture is treated with a mutagen and thousands of the resulting colonies are examined; when a mutant displaying significantly increased productivity is found, it serves as the starting point for a new round of mutagenesis and screening. In this way the organism's evolution is channeled in an unnatural direction until a strain is developed that produces an economic yield of the antibiotic.

Such work is slow and labor-intensive and its results are unpredictable, because antibiotic levels are strongly influenced not only by the genes of the pro-

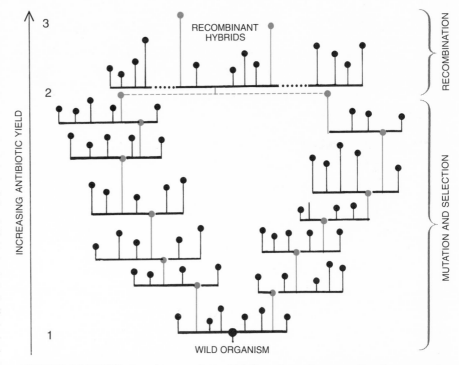

ANTIBIOTIC YIELD of a microorganism, which depends on a large number of genes, can be improved by mutation and selection followed by crossing, which recombines genes from two organisms. A wild strain is treated with a mutagen (*1*) and daughter colonies are screened; some of them will be better producers. The best strains (*color*) are selected and are again mutagenized. The process is repeated. The two best mutants (*2*), which in this case probably have six different mutations each and therefore differ from each other with respect to 12 genes or parts of genes, are crossed. Recombination can give rise to 2^{12} (almost 5,000) new genotypes, or sets of genetic information, many of which will lead to significantly higher antibiotic yields. Best producers (*3*) can again be subjected to recombination or to another round of mutation.

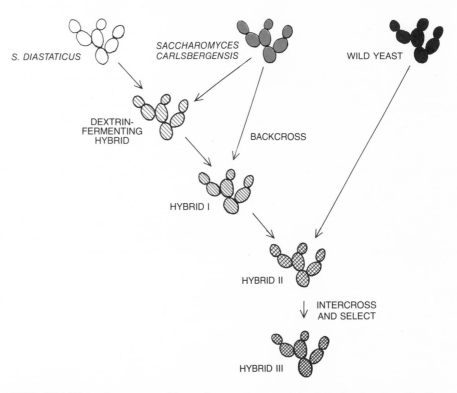

BEER YEAST *Saccharomyces carlsbergensis* converts sugars into alcohol and carbon dioxide, but it ferments only 81 percent of the sugars in the wort, or fermentation liquor. A yeast strain capable of fermenting all the sugar and thus making a very "light" beer suitable for diabetics is developed by crossing several yeast species to recombine genes encoding enzymes that convert different sugars. *S. diastaticus* ferments dextrins. Crossing it with *S. carlsbergensis* yields a dextrin-fermenting hybrid that makes unpalatable beer. Backcrossing hybrid several times with *S. carlsbergensis* yields hybrid I, which converts 90 percent of sugar and makes palatable beer. Crossing hybrid with wild yeast that ferments isomaltoses yields hybrid II, which converts 100 percent of sugar. Intercrossing strains of hybrid II yields improved strain (hybrid III).

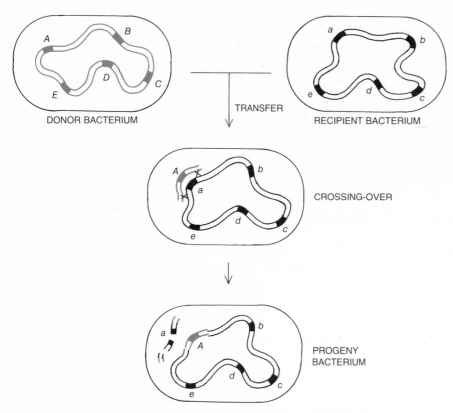

DONOR BACTERIUM

TRANSFER

RECIPIENT BACTERIUM

CROSSING-OVER

PROGENY
BACTERIUM

HOMOLOGOUS RECOMBINATION takes place in bacteria when a piece of the chromosome of a donor cell enters a recipient cell by one of several natural or artificially induced
processes. The donor DNA can pair with a homologous, or corresponding, region of the recipient chromosome, break and exchange segments in the process called crossing-over ($\times$ symbol),
yielding a new combination of the donor's and the recipient's genes. In the case of the five-
gene chromosome diagrammed here, one of 2^5, or 32, possible combinations of genes results:
donor gene *A* replaces gene *a* in chromosome; gene *a*, excluded from chromosome, breaks down.

ducing organism but also by culture
conditions. In the early stages of a yield-
improving program one can find advantageous mutants simply by testing the
yield of colonies growing on culture
plates, but eventually the improved
strains must be evaluated under conditions mimicking (as closely as possible)
those that will be encountered in commercial production in giant fermenters
containing 20,000 gallons (or more) of a
culture medium. In spite of these difficulties several antibiotic fermentations
now run at high productivity with cultures that have been developed through
20 or 30 selections spanning two decades or more.

Mutation alters a microorganism's
genes. Recombination, the other
basic approach to genetic programming,
rearranges genes or parts of genes and
brings together in an individual organism genetic information from two or
more organisms. Recombination results
from any of a wide variety of natural
processes and laboratory techniques.
Homologous recombination takes place
when bacterial or eukaryotic chromosomes that have similar DNA base sequences, brought together by some mating process, exchange corresponding
parts through the breaking and rejoining
of DNA. In the case of eukaryotes sexu-

al reproduction provides another shuf-
fling process, reassortment, in which the
sets of chromosomes derived from two
individuals are scrambled.

Homologous recombination is amazingly effective in producing new genotypes, or sets of genetic information. If
two individuals differ in *n* genes or parts
of genes, recombination between their
sets of genes will generate 2^n genotypes.
Matings between microorganisms of
two strains whose DNA differs in only a
dozen base pairs scattered among billions of base pairs can generate 2^{12}, or
almost 5,000, new genotypes. In most
cases the parent cells differ more than
that, and so astronomical numbers of
new combinations arise when they are
crossed. Although nearly all microorganisms are probably capable of exchang-
ing genes with related strains, there are
rather few instances in which the potential power of natural genetic recombination has been exploited to develop industrial cultures with desirable features
derived from more than one strain. Industrial yeasts provide some examples.

The simplest life cycle in yeasts is
one in which an individual strain is haploid rather than diploid. That is, it has
only one set of chromosomes carrying
among them a single complete set of
genes, rather than (as in the case of most
animals and plants) two sets of chromo-

somes carrying two copies of each gene.
A typical yeast cell can undergo sexual
reproduction, and thus genetic recombination, only when it encounters a related
strain of the opposite mating type, or
"sex." The two cells fuse to give rise to a
temporarily diploid cell, within which
haploid sexual spores are thereupon
formed containing different combinations of the genes of the parent cells.
Industrial yeasts may depart from this
simple pattern by having several sets of
chromosomes or by mating only inter-
mittently. The hybridization of different
strains—often strains of different species—has nonetheless played an important part in the development of industrial yeasts particularly well adapted to the
rapid production of bread by modern
factory methods, to increasing the alcohol content of liquors for distillation
and to brewing special beers from which
nearly all the soluble carbohydrates
have been removed.

Ordinarily only members of closely
related species mate successfully. The
natural barriers to recombination between dissimilar organisms can often be
broken down, however, by the preparation of protoplasts: bacterial or fungal
cells whose tough outer walls have been
removed to expose the thin cell membrane. Because cell membranes have
about the same composition in most
species, protoplasts of different species
can be induced to fuse and form a hybrid cell, exposing their genes to recombination.

Protoplast fusion may also prove to
be an effective technique for increasing
the frequency of intraspecies recombination in organisms in which natural
mating is a rare occurrence, as it is in
many species of the actinomycete *Streptomyces*. The protoplasts are prepared by
digesting the bacterial cell wall with the
enzyme lysozyme; the operation is carried out in a sugar solution whose osmotic pressure balances the pressure inside the cells, which would otherwise
burst the delicate cell membrane. The
fusion of protoplasts of two strains is
promoted by treatment with an agent
such as polyethylene glycol, and in the
resulting hybrid cell the DNA of the
parents may be recombined. The hybrid
protoplasts can then be induced to regenerate their cell wall to yield a normal
culture. Fusion is so efficient that cells of
two *Streptomyces* strains can be hybridized to give rise to a population in which
at least one cell in five has a new combination of genes. It should be possible in
this way to combine, in one step, groups
of mutated genes enhancing antibiotic
yield that have been accumulated labo-
riously in separate lines by successive
rounds of mutation and selection.

Whereas homologous recombination brings about an exchange of
corresponding stretches of DNA, other
forms of recombination add new DNA

to what is already possessed by a microorganism. One such process is the transfer of plasmids. These are small circular molecules of extrachromosomal DNA, found in bacteria and in some yeasts, that are capable of autonomous replication within a cell and are inherited by daughter cells. Plasmids often carry genes that give particular bacteria specialized properties. They can be transferred from one bacterial strain to an unrelated strain, and sometimes to a different species, to introduce totally new genetic properties. Some plasmids code for structures that cause bacteria to mate, and thereby promote their own transfer from cell to cell. Plasmids can also be carried from one bacterium to another by a bacteriophage, or bacterial virus. And naked plasmid DNA, having been liberated by the bursting (either natural or induced) of its host cell, can enter a new cell in the process called transformation, which can be greatly facilitated by various laboratory techniques.

An example of microbial breeding through the transfer of naturally occurring plasmids is the construction of a bacterium that is able to metabolize, and so to degrade, most of the major hydrocarbon components of petroleum. Many strains of *Pseudomonas putida* harbor a plasmid coding for enzymes that digest a single class of hydrocarbons. Four such plasmids are *OCT*, which digests octane, hexane and decane; *XYL* (digests both xylene and toluene); *CAM* (camphor), and *NAH* (naphthalene).

Two of the plasmids (*CAM* and *NAH*) bring about their own transfer by promoting mating between bacteria; the others do not, but they can be transferred if other plasmids that promote mating are supplied. By means of successive crosses a "superbug" has been created that carries the *XYL* and *NAH* plasmids as well as a hybrid plasmid derived by recombining parts of *CAM* and *OCT* (which are "incompatible" and cannot coexist as separate plasmids in the same bacterium). The multiplasmid bacterium grows rapidly on a diet of crude oil because it metabolizes much more of the hydrocarbon components than any one of the single-plasmid strains. Whether the superbug will be effective in cleaning up oil spills or scouring the holds of tankers remains to be seen, but it has already made legal history: its patent, the first ever granted a genetically manipulated microorganism, has been upheld by the U.S. Supreme Court.

There are other natural plasmids that carry genes coding for properties with industrial potential. In my laboratory at the John Innes Institute in Britain we have found that the synthesis of some of the antibiotics made by various *Streptomyces* species is controlled not by chromosomal DNA but by a plasmid; the

diversity of these antibiotics makes different ones effective for a broad range of applications, not only in human and veterinary medicine but also as animal-feed supplements and to control some plant diseases. The transfer into a single *Streptomyces* host of various plasmids that encode different antibiotics might make it possible for their enzymes to cooperate in the synthesis of new antibiotics with new properties.

The ability to isolate plasmid DNA from a culture and induce another culture to take it up is the basis of most recombinant-DNA manipulations. A gene or genes taken from an unrelated organism, or an artificially synthesized

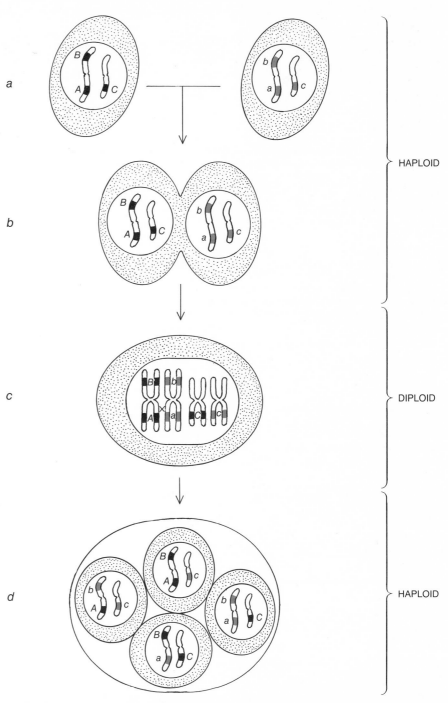

RECOMBINATION IN EUKARYOTES involves crossing-over between parts of chromosomes and the reassortment of chromosomes. A typical yeast is haploid during most of its life cycle, with a single set of 15 or more chromosomes; only two are shown here. Two cells of opposite mating types (*a*) can fuse (*b*); then the nuclei fuse to form a diploid nucleus with two complete sets of chromosomes. During meiosis (a phase of sexual reproduction) the chromosomes become double structures consisting of two chromatids (*c*). Homologous chromosomes pair and exchange parts of their chromatids by crossing-over. Then four haploid sexual spores are formed (*d*). Each spore can have a new combination of genes that were different (*black and color*) in parent cells: genes on the same chromosome (*A, a; B, b*) recombine by crossing-over and genes on different chromosomes are shuffled as members of chromosome pairs reassort.

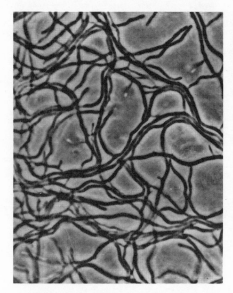
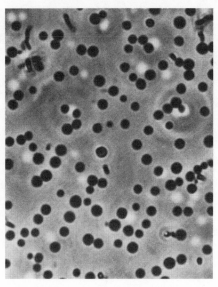

SPHERICAL PROTOPLASTS and the bacterial cells from which they were released are enlarged about 1,500 diameters in a micrograph made by Keith Chater and the author. A tough cell wall bounds the filamentous cells of the actinomycete *Streptomyces coelicolor* (*left*). The wall is digested with the enzyme lysozyme, leaving each cell enclosed only in its thin cell membrane; the protoplasts (*right*) are spherical because they are in a concentrated sugar solution whose osmotic pressure balances the pressure inside cells. Protoplasts can be induced to fuse.

gene, can be spliced into a plasmid; the plasmid is thereupon introduced into a new microbial host. The plasmid thus serves as a vector for genes that have no counterpart in the recipient organism and therefore could not be stably inherited in it through homologous recombination; such genes can now be passed on indefinitely through successive generations as the plasmid replicates. The

DNA of certain viruses can also serve as a vector, provided they can infect a microorganism and be inherited without killing the host. Many bacterial viruses (the "temperate" bacteriophages) can do this. So can some viruses that infect plants and animals, opening the way to the genetic engineering of plant and animal cells as well as microorganisms for industrial purposes.

Recombinant-DNA techniques can be applied in various ways for a number of different industrial purposes. The most widely known objective is the production by a microorganism of a protein it does not normally synthesize, such as an enzyme or a hormone. Here the idea is to transfer an individual gene coding for the desired product into a host microorganism and grow the organism in volume to yield the product. (So far the bacterium *Escherichia coli* has served almost exclusively as the host because effective vector plasmids, whose DNA has been mapped in fine detail, are available for this workhorse of molecular biology; eventually it may be preferable to choose other hosts better adapted to growth in large industrial fermenters.) A rather different objective of plasmid engineering is the genetic improvement of an existing industrial strain. Instead of introducing a brand-new genetic capability one can improve the efficiency of an existing strain by modifying its genetic information. Finally, recombinant-DNA techniques should make it possible to improve the precision of a traditional approach by bringing about the mutation of specific sites in particular genes, thereby overcoming the random nature of normal mutagenesis.

The basic methods of recombinant-DNA technology can be described in terms of their application to the production of a desired protein. Work over the past few years in laboratories around the world has made it relatively easy to cut giant DNA molecules, such as those in chromosomes, into a number of short fragments with the help of special enzymes, known as restriction endonucleases, that cleave DNA at specific base-pair sites. Some of these enzymes generate fragments with "sticky ends." Fragments carrying the gene to be transferred are inserted into plasmid (or bacteriophage) vectors cut open with the same enzyme and therefore having matching ends. The resulting recombinant plasmids are introduced into *E. coli* by transformation. Clones of bacteria (colonies of cells descended from a single parent cell) that harbor the plasmid can be selected because they can grow in the presence of an antibiotic, resistance to which is imparted by a gene on the plasmid. The presence of the desired foreign fragment on the recombinant plasmid can be recognized by means of sensitive tests for the desired protein product. Bacteria harboring the plasmid can then be grown into clones of billions of cells, each cell carrying a copy of the foreign gene.

It is possible in principle, although laborious in practice, to clone just about any desired gene, say the natural human gene coding for insulin. This can be done by "shotgun" cloning of human DNA: inserting into plasmids and then into *E. coli* a random mixture of gene-

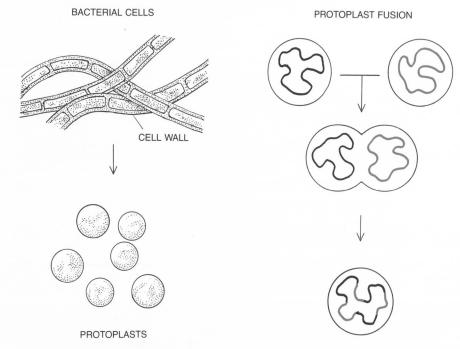

PROTOPLAST FUSION makes it possible to recombine the genes of two species that do not mate or to enhance recombination between strains of an organism such as *Streptomyces*, which mate infrequently. Protoplasts are formed from *Streptomyces* cells (*left*). Protoplasts of two strains are treated with polyethylene glycol (*right*). The protoplasts fuse, forming a new hybrid cell containing two chromosomes, segments of which recombine to yield new genotype.

size fragments of the total human DNA complement and then searching for the right transformed bacteria among a million or so clones. Such cells will not serve as a bacterial insulin factory, however, because bacteria cannot express eukaryotic genes incorporating introns.

An artificial, unsplit gene must be constructed.

One approach is to isolate messenger RNA extracted from the human pancreas cells that make insulin. These cells are rich in insulin messenger RNA,

from which the introns have already been spliced out. With the enzyme called reverse transcriptase one can make an artificial "copy DNA" carrying the uninterrupted genetic information for insulin, and the copy DNA can be cloned. Another approach (if the amino

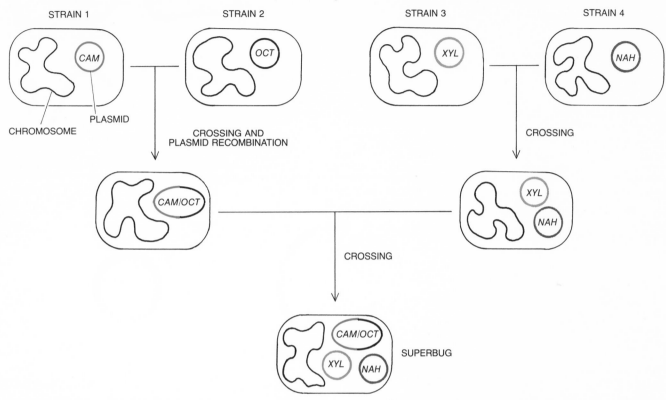

"SUPERBUG" that can metabolize the major hydrocarbons of petroleum is constructed by the manipulation of plasmids. The ability of four different strains of *Pseudomonas putida* to metabolize either the camphor (*CAM*), octane (*OCT*), xylene (*XYL*) or naphthalene (*NAH*) family of hydrocarbons depends on enzymes whose genes are not on the chromosome but on individual plasmids. A strain that can degrade all four families is constructed by successive crosses. Because *CAM* and *OCT* plasmids cannot coexist in one cell an extra step is required: hybrid plasmid is formed in which parts of the plasmids, carrying genes encoding necessary enzymes, are recombined.

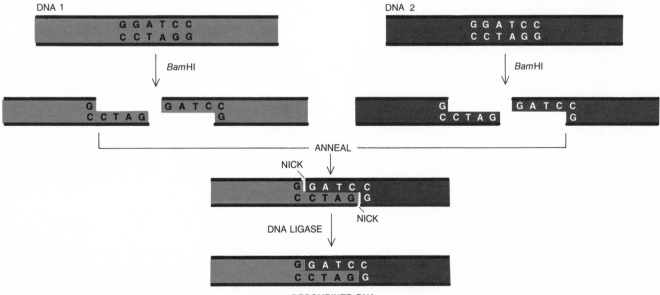

RESTRICTION ENDONUCLEASES, basic tools in recombinant-DNA technology, are bacterial enzymes that recognize particular short palindromic (rotationally symmetrical) sequences in a DNA strand and cut the strand at specific sites in those sequences. The enzyme *Bam*HI (named for its source, *Bacillus amylolyticus*) and many other endonucleases cleave DNA in a way that leaves single-strand protrusions, called "sticky ends" because their bases are complementary. Two DNA's from different sources (*color and gray*), both cleaved by *Bam*HI, can be "annealed" by base pairing. The "nicks" in backbones of annealed DNA are sealed by enzyme DNA ligase.

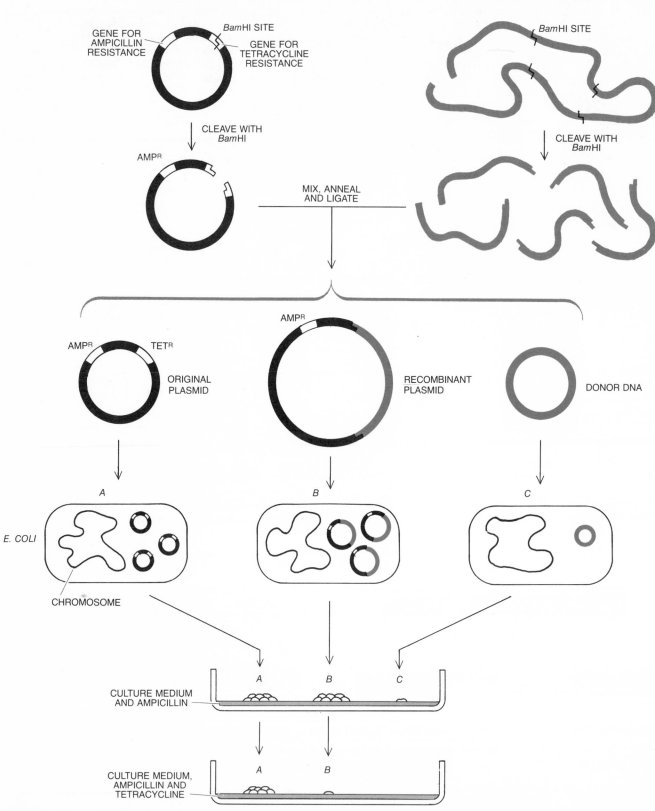

PLASMID FROM *ESCHERICHIA COLI*

DNA FROM DONOR ORGANISM

GENE FOR AMPICILLIN RESISTANCE

*Bam*HI SITE

GENE FOR TETRACYCLINE RESISTANCE

*Bam*HI SITE

CLEAVE WITH *Bam*HI

CLEAVE WITH *Bam*HI

AMP^R

MIX, ANNEAL AND LIGATE

AMP^R TET^R

AMP^R

ORIGINAL PLASMID

RECOMBINANT PLASMID

DONOR DNA

E. COLI

A

B

C

CHROMOSOME

A

B

C

CULTURE MEDIUM AND AMPICILLIN

A

B

CULTURE MEDIUM, AMPICILLIN AND TETRACYCLINE

FOREIGN GENE is introduced into the bacterium *Escherichia coli* **on a plasmid vector to yield a clone of bacteria carrying the gene. Here the plasmid carries genes for resistance to the antibiotics ampicillin and tetracycline. A plasmid treated with** *Bam*HI **is cleaved at one site, within the tetracycline-resistance gene; sticky ends are generated. DNA from a donor organism is cleaved by** *Bam*HI **into a number of fragments that have the same sticky ends. The cleaved vector and the donor DNA are mixed, annealed and ligated. A mixture of molecules is formed: recircularized plasmids (***left***), recombinant plasmids incorporating a segment of donor DNA (***center***) and circularized donor DNA (***right***).** *E. coli* **bacteria are transformed with** the mixture: they take up various molecules. If the transforming molecule is a plasmid, it replicates and is inherited by daughter bacteria, rendering them resistant to ampicillin. Such cells (*A, B*) are recognized because a sample of them will grow on a medium containing ampicillin, whereas untransformed cells, or cells transformed only by donor DNA (*C*), are killed. A "replica" of the ampicillin-resistant colonies is picked up on a disk of velvet and is transferred to a medium containing tetracycline as well as ampicillin; now any cells harboring recombinant plasmids (*B*) are killed because the tetracycline-resistance gene on the plasmids has been disrupted. Culture that gave rise to colony *B* is tested to identify clones carrying particular donor genes.

acid sequence of a protein is known, as it is in the case of human insulin) is to synthesize an artificial gene by assembling the appropriate DNA nucleotides according to the genetic code. This has been done for some short proteins; the advent of "gene machines" that can synthesize specified stretches of DNA automatically may make the method more generally feasible.

Neither a copy-DNA gene nor an artificially synthesized gene carries the appropriate promoter and other control signals needed for expression in bacteria. It is necessary, therefore, to splice the artificial gene into the vector DNA at a point where the bacterial control signals on the vector will lead to expression of the artificial gene. That is not the end of the task. Bacteria often digest foreign proteins, so that the host cell may have to be modified to keep it from destroying its own novel product; a mutant host may have to be selected that has lost the enzymes for protein digestion. The host must also be selected or modified to grow well in spite of having to accumulate or secrete large quantities of a protein for which it has no use. The commercial stimulus for developing successful fermentations producing such therapeutic proteins as human insulin, growth hormone or the antiviral agent interferon is so great that these various obstacles to volume production are likely to be overcome very soon.

It should even be possible to program microorganisms to make proteins that do not occur naturally in any organism. The recent successful cloning of copy-DNA genes for human interferon has revealed that there is a surprisingly large number of interferons, which differ not only in their amino acid sequence but also in their properties. One might make brand-new members of the interferon family by an artificial equivalent of homologous recombination: by splicing together parts of two genes, isolated from two *E. coli* clones, coding for different natural interferons. The hybrid gene would then be reintroduced into the bacterial host. This approach could provide a wealth of new molecules to be tested for their therapeutic value. Later, when more is known about the relations between the architecture of proteins and their biological properties, it should even be possible to design entirely new proteins and manufacture them by synthesizing and cloning artificial genes.

The genetic improvement of an existing strain is appropriate in the case of substances, such as antibiotics and alkaloids, that are not directly encoded by genes but instead are synthesized by pathways controlled by a number of gene products. Rather than attempting the daunting task of transplanting the entire set of genes for such products into a wholly foreign host unaccustomed to

expressing them, one can modify the genetic information of an existing industrial strain. One might, for example, increase the flow through a metabolic bottleneck by adding duplicate copies of a gene, or provide a microorganism with a new enzyme that can modify a natural

metabolite to make a wanted product. Cloning systems are being developed for such purposes.

An example of the reprogramming by genetic engineering of a good industrial microorganism to make it even better is that of a strain of the bacterium *Meth-*

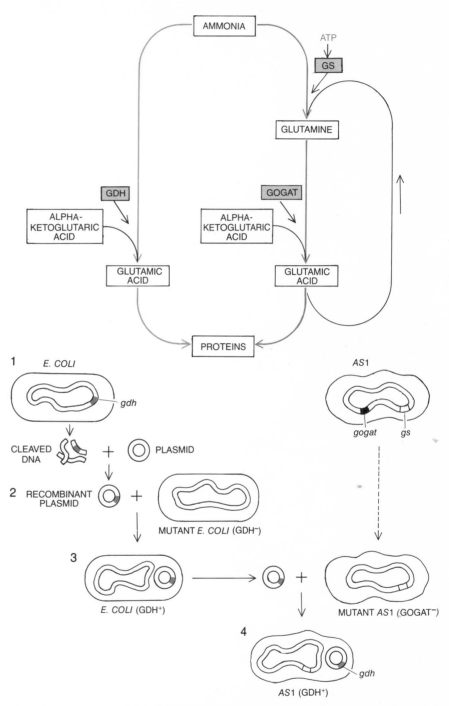

GENETIC ENGINEERING substitutes an energy-conserving nitrogen pathway (*light color*) for the similar but energy-consuming pathway (*dark color*) in an industrial microorganism, a strain of *Methylophilus methylotrophus* called *AS*1. Both pathways take nitrogen from ammonia and pass it on to amino acids to form proteins, but the step to glutamic acid catalyzed by the enzyme GDH (in *E. coli* and some other species) does not require energy delivered by ATP, whereas one of the two steps catalyzed by the enzymes GS and GOGAT (in *AS*1) does. By means of recombinant-DNA techniques the gene encoding GDH is isolated from *E. coli* and spliced into a plasmid (*1*). Mutant *E. coli* lacking GDH are transformed with the recombinant plasmid (*2*); transformed cells are recognized by their ability to obtain nitrogen from ammonia. Recombinant plasmids are reisolated and introduced into mutant *AS*1 cells lacking gene for GOGAT (*3*). Transformed *AS*1 cells (*4*) synthesize GDH and therefore grow more efficiently.

ylophilus methylotrophus designated *AS*1. The cells are grown in giant fermenters and then are harvested and dried to yield a protein-rich animal-feed supplement. *AS*1 depends on methanol as a source of carbon and on ammonia as a source of nitrogen. The pathway by which it builds ammonia into proteins by way of glutamic acid starts with two reactions catalyzed by the enzymes glutamine synthetase (GS) and glutamate synthase (GOGAT). The pathway is wasteful of energy because the step catalyzed by GS is driven by the cellular energy transducer adenosine triphosphate (ATP). In some other bacteria, including *E. coli*, there is a pathway that depends on a different enzyme, glutamate dehydrogenase (GDH), to make glutamic acid, and it requires less energy.

An energy-conserving strain of *AS*1 has been developed. First an auxotrophic mutant of *AS*1 was isolated that lacked GOGAT; it could not convert ammonia into proteins. Then the *E. coli* gene for GDH was cloned by introducing fragments of normal *E. coli* chromosomal DNA, spliced into a plasmid vector, into an *E. coli* mutant strain that lacked GDH; the desired clone could be recognized because, unlike the mutant *E. coli*, it was able to utilize ammonia. The plasmid vector had been chosen for its ability to be transferred (by mating or by transformation) into bacteria of many different species, including *AS*1. And so, in the final step, it was possible to transfer the cloned GDH gene into the *AS*1 auxotroph that lacked GOGAT. The resulting strain, equipped with GDH instead of GOGAT, does indeed grow with somewhat less expenditure of energy. Even a modest increase in energy efficiency is valuable when thousands of tons of a commodity are being produced.

A particularly exciting application of recombinant-DNA technology is site-directed mutation. The randomness of spontaneous or even induced mutagenesis has made it hard to find mutant organisms with alterations at specific sites in their DNA. That is not critical when one wants to develop an auxotrophic mutant lacking a particular enzyme, because the target for mutagenesis is rather large: a change of any one of hundreds of base pairs in a gene can inactivate the gene. It is much harder, however, to alter a specific part of a gene in a way calculated to improve its performance, say to change a particular base pair in a promoter region to increase the rate of transcription. Now that a gene can be isolated from a clone its DNA sequence can be altered by specific chemical treatment outside the cell. Then the gene can be reintroduced into the host and homologous recombination can be relied on to exchange the new gene for the normal one.

Industrial microbial genetics has now come of age. A range of techniques for genetic programming is available that could not have been envisioned only a few years ago. They include site-directed mutagenesis to help overcome the intrinsic randomness of procedures based on the isolation of mutants, protoplast fusion to increase the power of natural recombination, and an entire battery of recombinant-DNA manipulations for transplanting natural genes and even making new ones. The judicious application of these techniques, alone and in combination, should enormously expand man's ability to understand the amazing biochemical versatility of microorganisms, to enhance it and to channel it wisely.

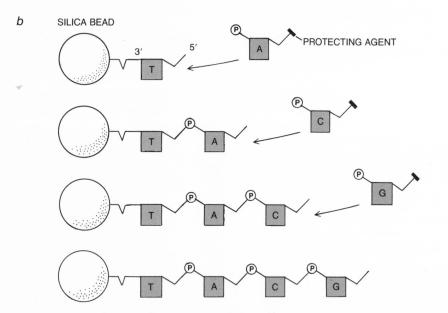

"GENE MACHINES," several versions of which are now available, synthesize specified short sequences of single-strand DNA automatically and very quickly under the control of a microprocessor. A version made by Bio Logicals, a Canadian company, is diagrammed (*a*). The desired sequence of bases is entered on a keyboard. The microprocessor opens valves that allow successive nucleotides and the reagents and solvents needed at each step to be pumped through the synthesizer column. The column is packed with tiny silica beads (about the consistency of fine sand); each bead serves as a solid support on which the DNA molecules are assembled. To make a given sequence (*b*) a column is used in which many thousands of copies of the nucleoside (base plus sugar) that is to be at the so-called 3′ end of the sequence (a *T* in this case) have already been fixed to each bead, leaving the nucleoside's 5′ side free. The microprocessor pumps millions of copies of the next nucleotide (*A*), with its 5′ side protected against unwanted reactions, into the column, and the *A*'s bind to the *T*'s. The protecting agent is removed, leaving the 5′ side free to accept the next nucleotide (*C*). In this way chains of about 40 nucleotides have been synthesized at the rate of about one nucleotide every 30 minutes. The completed chains are cleaved from the beads and are eluted into the collector.

4

The Microbiological Production
of Food and Drink

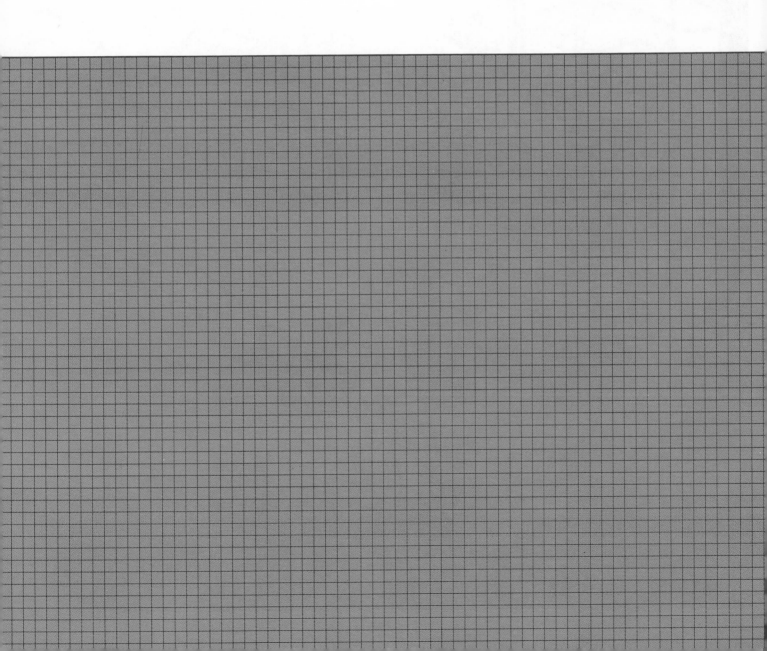

The Microbiological Production of Food and Drink

BY ANTHONY H. ROSE

Beer, wine, bread and cheese have been made by micro-organisms since Neolithic times. To them have been added spirits, yogurt, pickles, sauerkraut, Oriental fermented foods and today single-cell protein

Microorganisms were improving and spoiling the food and drink of human beings long before anyone realized that microorganisms exist. In time, but still without knowing what was happening biologically, people learned to encourage and exploit the fermentative action of microorganisms in the making of such things as cheese and beer. Today, with microbial activity fairly well understood, fermented foods and beverages constitute a large and important sector of the food industry. With the advent of the genetic-programming techniques David A. Hopwood discusses in the preceding article one can foresee large-scale advances in the quality and precision of the microbiological production of food and drink.

Milk was probably one of the first agricultural products. Since milk is quickly infected by bacteria, which sour it by converting its sugar (lactose) into lactic acid, it is likely that cheese was one of the first fermented foods. Over the millenniums this spontaneous process was gradually exploited and developed to make cheese and similar products. The manufacture of cultured dairy products is now second (in sales) only to the production of alcoholic beverages among the industries that rely on microbiological processes.

Cheesemaking basically calls for adding a starter culture of bacteria to the milk and letting the mixture incubate for a while. Then a proteolytic (protein-digesting) enzyme is added to coagulate the solids in the souring milk. Traditionally calf rennet, obtained from the fourth stomach of the unweaned calf, was the source of the enzyme, but it is gradually being replaced by microbial enzymes. The coagulated curd is separated from the whey, pressed to squeeze out some of the water and wrapped in cloth to dry. With some cheeses the growth of microorganisms on the outside of the cheese is encouraged during the curing process.

By about a century ago the art had advanced to the point where cheesemakers could stop relying on the spontaneous infection of milk by bacteria and could instead exploit one or more of several species of bacteria specifically cultured as cheese starters. The nature of the bacteria serving as the starter is one of several factors contributing to the enormous variety of cheeses. Other factors include the temperature of manufacture and the presence or absence of a secondary microbial flora on the cheese.

Soft cheeses have a high water content (from 50 to 80 percent) and are classified as ripened or unripened. A ripened soft cheese is a finished product as it comes from the initial processing steps; cottage cheese is an example. In an unripened soft cheese such as Camembert or Brie the growth of yeasts and species of the fungus *Penicillium* on the surface of the cheese is encouraged. If the cheese is to be semihard, it is cooked briefly to lower the moisture content of the curd to about 45 percent, thus making the curd firmer. Some varieties (Caerphilly for one) have a flavor like fresh curd; others (such as Limburger) are soaked in brine, which causes a surface flora of yeasts and bacteria to develop.

Hard cheeses, in which the water content is 40 percent or less, may have a simple bacterial flora, as Cheddar cheese does. Other hard cheeses differ in that the curd is inoculated with spores of mold (usually *Penicillium roqueforti*) that germinate when the curd is spiked to admit air. The growth of the mold in the cheese generates the flavor and aroma compounds that are characteristic of the individual cheese. Stilton, Danish blue, Roquefort and Gorgonzola are examples. A third class of hard cheeses, which include Gruyère, differs in that bacteria producing propionic acid are added to the starter mix. These bacteria, such as *Propionibacterium shermanii,* not only give the cheese a characteristic flavor but also, by generating carbon dioxide gas, give rise to the holes typical of such cheeses. Swiss mountain cheeses are in this class, but they are allowed to develop a surface flora of yeasts and bacteria that adds to their flavor and aroma.

Another type of fermented milk product differs from cheese in that it is liquid or semiliquid. The most popular members of the class are the yogurts. Yogurt is made by fermenting whole milk with a symbiotic mixture of two lactic acid bacteria, *Streptococcus thermophilus* and *Lactobacillus bulgaricus.* The fermentation is done at a temperature of about 40 degrees Celsius (104 degrees Fahrenheit). The characteristic flavor of a yogurt is attributable to lactic acid, made from the lactose in the milk, and to acetaldehyde; both are formed mainly by the *L. bulgaricus* bacteria. Because many people do not care for the tartness and acetaldehyde flavor of fresh yogurt the product is often flavored with fruit or fruit essences; more than 90 percent of the 500,000 pounds of yogurt produced in the U.S. each year are flavored in this way.

Related to the yogurts are a number of other fermented dairy products. Sour cream, for example, is made by souring pasteurized cream with lactic acid bacteria. Buttermilk is made by fermenting skimmed or partly skimmed pasteurized milk with a mixture of lactic acid bacte-

COPPER BREW KETTLES at the Pabst Brewing Co. plant in Milwaukee appear in the photograph on the opposite page. They serve in the boiling stage of brewing beer. The material boiled is wort, a water extract of germinated barley supplemented with hops to give flavor to the beer. Later the wort is put in tanks and fermented by a strain of the yeast *Saccharomyces cerevisiae.* Boiling in a brew kettle is part of an ancient technology that has been modernized, as is indicated by the control panel above the stained-glass window. The figure in the window is Gambrinus Rex, or King Gambrinus, a mythological ruler said to be the patron saint of beer.

ria and related species. Among the more exotic products in this category are Bulgarian milk, kefir and koumis, which are popular in Slavic countries, and *vilia,* which is popular in Finland.

An ancient process that relies on microbial activity is the preservation of vegetables. It was in service long before the advent of canning and freezing and is still practiced on a commercial scale in several countries. Cabbage, olives and cucumbers in particular are preserved by a combination of brine treatment and fermentation. The vegetable is treated in a succession of brines containing different concentrations of salt. The final concentration is as low as 2 percent for cabbage and as high as 18 percent for olives. (Some pretreatment of the vegetable may be necessary. Olives, for example, contain an extremely

bitter phenolic glucoside called oleuropein and need to be treated with a dilute solution of sodium hydroxide before brining to remove the bitterness.)

While the vegetables are in the brine they are subjected to the activity of a succession of microorganisms. The first step is the growth of the predominantly aerobic microbial flora that was on the surfaces of the vegetables before brining. Soon, however, the originally small numbers of lactic acid bacteria take over and, together with certain fermentative yeasts, including species of *Saccharomyces* and *Torulopsis,* carry out a fermentation that results in the production of lactic and acetic acids. Later the yeasts take over from the lactic acid bacteria. Fermentation ends when all the fermentable carbohydrates have been used up, although other species of yeast (mainly of the genera *Pichia, Debaro-*

myces and *Candida*) continue to grow as a film on the surface of the brine. In order to avoid having to rely on the flora of bacteria and yeasts naturally present in the vegetables and the brine, efforts have been made with some success to introduce cultures of the starter type, particularly of lactic acid bacteria, to better control the fermentations.

Whereas milk and vegetables are fermented primarily to preserve the nutrients of these basic foods, the growth of microorganisms is encouraged in other traditional types of fermented foods mainly to improve the taste and flavor of the product. Simultaneously the growing microorganisms increase the protein content of the food. Fermented foods of this class, which originated in the Orient, have fish or plant material (particularly soybeans) as the starting material. Fermented fish products are still largely limited to local consumption in Oriental countries, but soybean fermentations have gone farther afield, particularly among Oriental communities in North America.

Typical of these foods are the tempehs. A well-made tempeh consists of a compact cake of plant material completely covered and penetrated by white mold mycelia of species of the fungus genus *Rhizopus.* A word following "tempeh" designates the nature of the plant material that has been fermented. For example, *tempeh kedele* is made from *kedele,* the Indonesian word for soybeans. *Tempeh bongkreg katjang* is made from peanuts, *tempeh enthoe* from coconuts.

The plant material is soaked in water, dehulled, boiled or steamed and drained of excess water. Then raw tempeh from a previous batch is mixed in to supply spores of the *Rhizopus* mold. The mash is placed in trays, or in banana leaves when the tempeh is made in villages, and left until the mold has penetrated it sufficiently. Tempeh is usually not eaten raw but is deep-fried in coconut oil or cooked in some other way. Containing as much as 40 percent protein, tempehs are widely consumed in Indonesia.

Slightly different procedures are employed to make indigenous fermented foods in other countries. In Japan *natto* is the name given to the product that results when whole soybeans are fermented with the mold *Aspergillus oryzae.* A traditional Chinese food is *sufu,* a soft, cheeselike product made by fermenting soybean curd with a variety of molds, principally species of *Mucor.* Another variation on the theme is *ang-kak,* which originated in China. It is made by fermenting rice with the mold *Monascus purpureus.* Here the objective is not to alter the flavor of the rice but merely to color it red.

Soy sauce is a widely known product of the fermentation of soybeans. It was originally brewed in China many centuries ago and later introduced into other

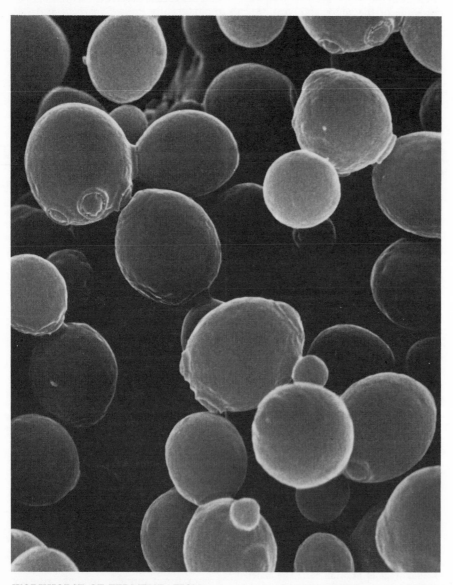

WORKHORSE OF FERMENTATION is the yeast *Saccharomyces cerevisiae,* **cells of which appear in this scanning electron micrograph made by Alastair T. Pringle of the University of California at Los Angeles. Each cell is about 10 micrometers in diameter. Strains of the yeast serve to raise bread and to make alcoholic beverages. As the cells ferment sugars they evolve carbon dioxide and also make alcohol; in dough the carbon dioxide forms the holes of bread. Several of the cells are budding (beginning to reproduce) and several budding scars are visible.**

CHEESE	ORIGIN	MICROORGANISM		
SOFT, UNRIPENED				
COTTAGE CREAM NEUFCHATEL	CENTRAL EUROPE ? U.S. FRANCE	*Streptococcus lactis* *Streptococcus cremoris* *Streptococcus diacetilactis*	*Leuconostoc citrovorum*	
SOFT, RIPENED				
BRIE	FRANCE	*Streptococcus lactis* *Streptococcus cremoris*	*Penicillium camemberti* *Penicillium candidum*	*Brevibacterium linens*
CAMEMBERT	FRANCE	*Streptococcus lactis* *Streptococcus cremoris*	*Penicillium camemberti* *Penicillium candidum*	
LIMBURGER	BELGIUM	*Streptococcus lactis* *Streptococcus cremoris*	*Brevibacterium linens*	
SEMISOFT, RIPENED				
ASIAGO	ITALY	*Streptococcus lactis* *Streptococcus cremoris* *Streptococcus thermophilus*	*Lactobacillus bulgaricus*	
BLUE	FRANCE	*Streptococcus lactis* *Streptococcus cremoris*	*Penicillium roqueforti* or *Penicillium glaucum*	
BRICK	U.S.	*Streptococcus lactis* *Streptococcus cremoris*	*Brevibacterium linens*	
GORGONZOLA	ITALY	*Streptococcus lactis* *Streptococcus cremoris*	*Penicillium roqueforti* or *Penicillium glaucum*	
MONTEREY	U.S.	*Streptococcus lactis* *Streptococcus cremoris*		
MUENSTER	GERMANY	*Streptococcus lactis* *Streptococcus cremoris*	*Brevibacterium linens*	
ROQUEFORT	FRANCE	*Streptococcus lactis* *Streptococcus cremoris*	*Penicillium roqueforti* or *Penicillium glaucum*	
HARD, RIPENED				
CHEDDAR	BRITAIN	*Streptococcus lactis* *Streptococcus cremoris* *Streptococcus durans*	*Lactobacillus casei*	
COLBY	U.S.	*Streptococcus lactis* *Streptococcus cremoris* *Streptococcus durans*	*Lactobacillus casei*	
EDAM	NETHERLANDS	*Streptococcus lactis* *Streptococcus cremoris*		
GOUDA	NETHERLANDS	*Streptococcus lactis* *Streptococcus cremoris*		
GRUYERE	SWITZERLAND	*Streptococcus lactis* *Streptococcus thermophilus*	*Lactobacillus helveticus*	*Propionibacterium shermanii* or *Lactobacillus bulgaricus* and Propionic bacterium *freudenreichi*
STILTON	BRITAIN	*Streptococcus lactis* *Streptococcus cremoris*	*Penicillium roqueforti* or *Penicillium glaucum*	
SWISS	SWITZERLAND	*Streptococcus lactis* *Streptococcus thermophilus*	*Lactobacillus helveticus*	*Propionibacterium shermanii* or *Lactobacillus bulgaricus* and Propionic bacterium *freudenreichi*
VERY HARD, RIPENED				
PARMESAN	ITALY	*Streptococcus lactis* *Streptococcus cremoris* *Streptococcus thermophilus*	*Lactobacillus bulgaricus*	
ROMANO	ITALY	*Lactobacillus bulgaricus*	*Streptococcus thermophilus*	
PASTA FILATA (PLASTIC CURD)				
MOZZARELLA	ITALY	*Streptococcus lactis* *Streptococcus thermophilus*	*Lactobacillus bulgaricus*	
PROVOLONE	ITALY	*Lactobacillus bulgaricus*		

WELL-KNOWN CHEESES are charted according to their classification as ripened or unripened. A ripened cheese is a finished product when it emerges from the initial fermentation. In unripened cheeses the growth of yeasts and species of the fungus *Penicillium* on the surface of the cheese is encouraged after the initial processing. This list is far from exhaustive; the variety of cheeses is enormous.

Oriental countries, particularly Japan, which is now the main manufacturer. It is made by fermenting a salted mixture of soybeans and wheat with the mold *Aspergillus oryzae* to yield a mixture called *koji,* which is put in a vessel with an equal amount of salt solution to make a mash known as *moromi.* The *moromi* is fermented in large tanks for from eight to 12 months, with a certain amount of agitation and preferably at a low temperature. The microorganisms chiefly responsible for the fermentation (the bacterium *Pediococcus soyae,* the yeast *Saccharomyces rouxii* and species of the yeast genus *Torulopsis*) originate in the *moromi.* Occasionally starter cultures of these microorganisms are added to the *moromi,* but in either process the metabolism of the microorganisms enriches the *moromi* with lactic acid and

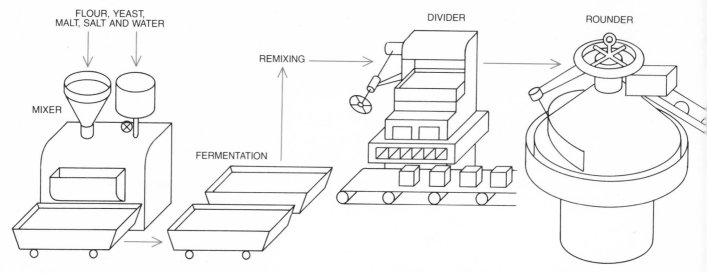

MANUFACTURE OF BREAD is portrayed in this flow chart of the major steps. A "sponge," or starting mixture, containing only part of the flour that will eventually go into the bread, is kneaded in a mixer for several minutes and then fermented for several hours. The rest of the flour is added and the dough is remixed. The divider cuts the dough into loaf-size pieces, which are further shaped by the rounder.

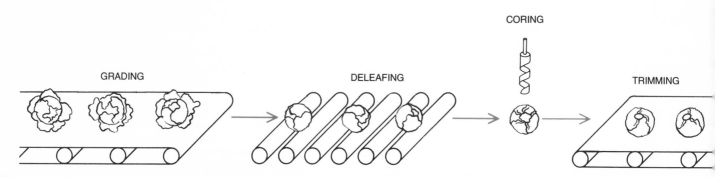

PRESERVATION OF VEGETABLES is an application of fermentation that long predates canning and freezing and is still widely practiced on a commercial scale. The process is depicted here for the preservation of cabbage as sauerkraut by the dry-salt method. In this

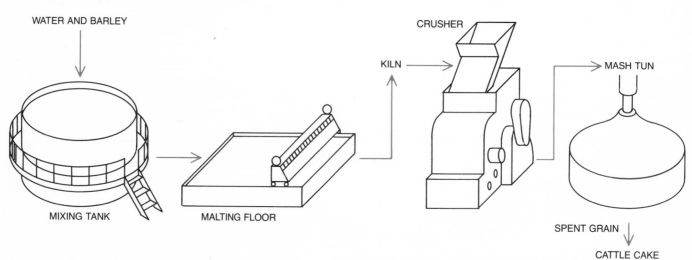

PRODUCTION OF BEER starts with the malting of barley, in which the grain is induced to sprout briefly to produce enzymes that will catalyze the breakdown of starch. The malt is ground and mixed with warm water (and often with other cereals such as corn) before going to the mash tun, where over a period of a few hours enzymes break down the long chains of starch into smaller molecules of carbo-

other acids and with ethanol. When the fermentation is complete, the *moromi* is pressed and the extruded soy sauce is packaged. The cake that remains is often fed to animals.

The discovery of flour and hence of breadmaking is thought to have been made very early in human development, probably in Egypt. The first breads were flat ones made by baking a mixture of flour and water. Precisely when dough was first leavened is not known. The main effect of leavening is to increase the volume of dough as a result of the breakdown of sugars by yeast to form bubbles of carbon dioxide. The bubbles are trapped in the dough, and when it is baked, they give leavened bread its characteristic honeycomb texture. Leavening may have come about as a result of the spontaneous growth of

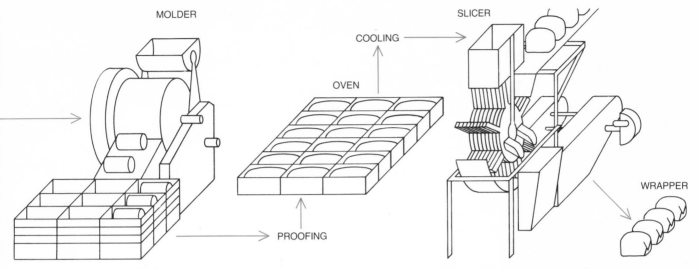

At this stage the dough is rubbery; it is put through an intermediate proofing stage (not shown), where it rises and changes in structure and is therefore easier to mold. The molder shapes the pieces of dough into cylinders that are put in baking pans. In a final proofing stage the dough ferments further before being put in the oven. After baking for 20 minutes the loaves of bread are cooled, sliced and wrapped.

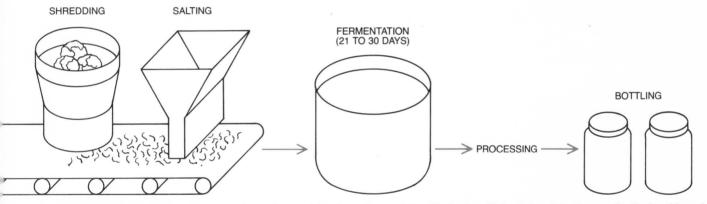

method a brine is generated by osmotic gradients arising from the interaction of the salt and the natural fluids of the cabbage. In the brine the lactic acid bacteria originating on the fresh cabbage become the dominant species in the extended process of fermentation.

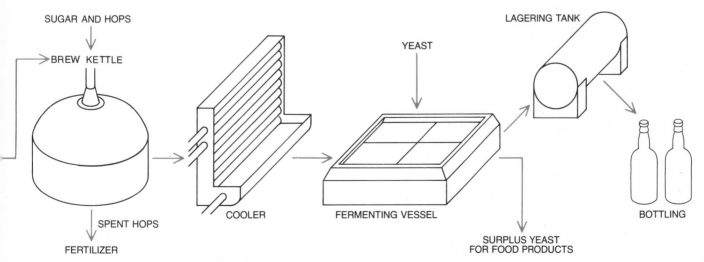

hydrate. The aqueous extract called wort is separated from the mix and boiled with hops in a brew kettle. The boiling extracts flavors from the hops and stops the enzyme action in the wort. The hops are removed and the wort is put in a fermenting vessel, where it is pitched, or seeded, with yeast. After fermentation the beer may go to a lagering tank to mature, following which it is pasteurized and bottled.

yeast in the mixture of flour and water or of the addition of fermenting beer to the dough.

Today breadmaking and the large-scale cultivation of yeast that is associated with it constitute one of the most sophisticated branches of industrial microbiology. Although flat breads are still made in many parts of the world, most bread in the developed countries is made by mixing flour (usually wheat flour) with water and smaller proportions of yeast, salt, sugar and shortening. After it is mixed or kneaded the dough is allowed to ferment at a temperature of about 25 degrees C. (higher in recently developed processes). During this time the yeast, a strain of *Saccharomyces cerevisiae,* breaks down sugars in the dough into a mixture of alcohol and carbon dioxide gas, bubbles of which become fixed in the dough. This is fermentation. When the dough is baked after the period of fermentation, the alcohol is driven off, but the bubbles of carbon dioxide remain to give texture to the bread. Some sugars are available to the yeast immediately, including added ones such as sucrose and cane sugar. They are supplemented by sugars liberated from the

starch of the cereal grain by two enzymes, alpha-amylase and beta-amylase, that are constituents of the flour and are activated by water. The sugars include maltose and glucose. Maltose is usually fermented by the yeast toward the end of the fermentation process, when the other sugars have been almost used up.

Although the main function of the yeast in bakery fermentations is to raise the dough, it also has other effects. One effect is to change the structure and texture of the dough, which it does by modifying the structure of gluten, the principal wheat protein, as the dough is stretched mechanically. Moreover, by excreting compounds such as cysteine and glutathione, the yeast may alter the structure of gluten by breaking intramolecular disulfide (S–S) bonds. Products of the fermentation by yeast also modify the flavor of the baked dough and increase its nutritive value to a limited extent.

Over the past 25 years or so this method of breadmaking, which is often called the bulk-fermentation process, has been modified to enhance the possibilities of handling the dough rapidly by

machine. All the rapid methods seek to produce baked dough at a faster rate than bulk fermentation does, and to this end the fermentation is done at a higher temperature (usually about 35 degrees C.) and the dough contains a higher proportion of yeast, which may also be a strain with greater fermentative activity. In addition the dough is subjected to intense mechanical mixing, which has an effect on the structure of the dough.

Modern breadmaking could not be the efficient industry it is without the associated industry manufacturing baker's yeast. Until the middle of the 19th century the yeast that went into bread dough was barm, the residual yeast from the brewing of beer. Barm proved to be unreliable, however, as the volume of breadmaking increased, and specialized plants were therefore built for the production of baker's yeast. In such plants specially selected strains of *Saccharomyces cerevisiae* are grown in highly aerated conditions in a nutrient medium based on molasses. The bulk production of baker's yeast must be done under closely controlled conditions to ensure the constant fermenting ability that bakeries require day after day.

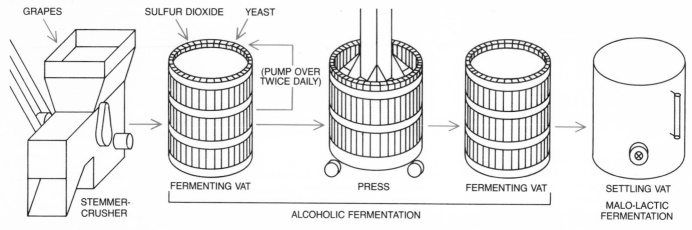

MAKING OF WINE is portrayed in a generalized process that in actuality differs somewhat for red wine and white wine. Here the wine is made in batches, which is by far the commonest way. Some of the cheaper wines are manufactured by a process of continuous fermen-

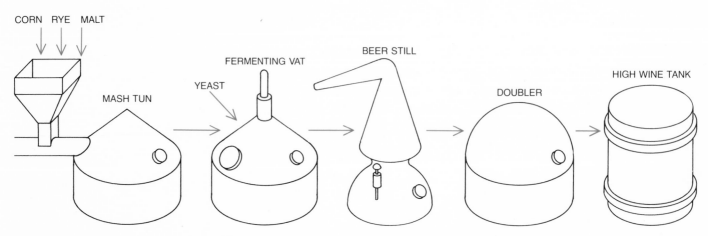

DISTILLED SPIRITS such as whiskey are made in a process that is much like the brewing of beer. It is depicted here for bourbon. Grains of corn are mixed with smaller amounts of rye and malted barley, crushed and mixed with warm water. The wort that emerges from the mash tun is transferred to a fermenter and pitched with yeast. After fermentation the beer is conveyed to a unit consisting of

The manufacture of alcoholic beverages also exploits the fermentation of sugars by *S. cerevisiae,* but here the main requirement is the alcohol rather than the carbon dioxide. Alcoholic beverages are grouped in three categories: the wines and beers, which are made by fermenting with yeast the juice of a fruit or a sugary extract of grain; the fortified wines, in which brandy is added to the wine, and the spirits, which are made by distilling wines or beers.

Any solution of the sugary substances of grain that is allowed to stand will soon become infected by microorganisms. Archaeological evidence shows that the fermentation of grain extracts was already an advanced art more than 6,000 years ago. The beers so made not only tasted better than water but also were safer to drink, since pathogenic organisms cannot grow in beer because of its acidity and its content of antimicrobial compounds derived from hops. The worldwide production of beer is now about 700 million hectoliters (18.5 billion gallons) per year, with the per capita consumption highest in West Germany and Australia.

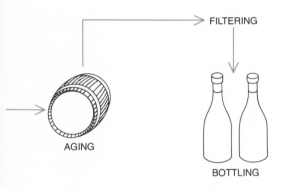

tation. Grape juice is fed steadily into a fermenting stage and wine is steadily removed.

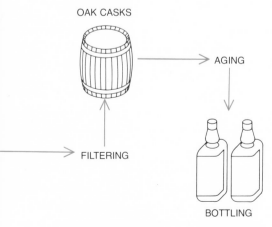

a beer still and doubler. The condensate is collected in a high wine tank and then matured for several years in oak casks before bottling.

Most beer is made from barley, although small amounts are made from other cereal grains. The grains of barley are first malted, that is, allowed to germinate for a short time. The main objective of the malting is to produce enzymes in the grain that (either during malting or later) catalyze the breakdown of starch. The malted barley is then crushed and mixed with water at a temperature of up to 67 degrees C. Within a few hours enzymes in the mash break down the long chains of starch into smaller carbohydrate molecules and also break down other long-chain molecules such as proteins.

The aqueous extract, which is called malt wort, is separated from the spent grains and boiled, traditionally with hops to give flavor to the final beer. The boiling of the wort not only extracts flavor compounds from the hops but also stops further enzyme action in the wort and precipitates protein from it. Now the hopped wort is pitched, or seeded, with a strain of *S. cerevisiae.* The main action of the yeast is to convert the sugars in the wort into alcohol and carbon dioxide. (There is also a fivefold increase in the amount of yeast during the fermentation.) Other quantitatively minor products of yeast metabolism have a strong effect on the flavor of the final beer. They include higher alcohols such as amyl, isoamyl and phenylethyl alcohol, which are present in beer at concentrations on the order of milligrams per liter. Other important flavor compounds formed by the yeast are short-chain acids such as acetic and butyric acids and esters of them. At the end of fermentation the yeast is separated from the beer, which is then allowed to mature for an appropriate period. After filtration, pasteurization and possibly other steps the beer is ready to be packaged and sold.

Traditionally two types of yeast are employed in brewing beer. The majority of beers are lagers, which by tradition are made with a yeast that settles to the bottom of the fermentation tank during fermentation. These bottom-fermentation yeasts were first isolated in pure culture about 100 years ago by the Danish botanist Emil Christian Hansen, working at the Carlsberg Institute in Copenhagen, and have been named *Saccharomyces carlsbergensis.* In Britain and in parts of Europe and North America the yeast employed in brewing beer rises to the surface during fermentation. Top-fermentation yeasts are classified as strains of *S. cerevisiae.* Taxonomists do not now distinguish these yeasts as separate species, although the two names continue to be used in brewing. The strains of yeast popular in breweries have been chosen by largely empirical means over centuries of brewing, but attempts are now being made to tailor the genetic makeup of brewing yeasts to the

requirements of the individual brewer.

The technology of wine making is much simpler. Until recently the process had changed little over the 5,000 years that wine has been made. Red or white grapes from selected varieties of the vine are collected and crushed to express the juice. Until recently the grape juice was allowed to ferment spontaneously by way of microorganisms present on the surface of the freshly picked grapes. The natural flora of the skins of grapes includes several different yeasts, some of genera other than *Saccharomyces.* Many of the yeasts responsible for the first part of the fermentation are later killed off by the alcohol released when strains of *S. cerevisiae* ferment sugars in the juice. After fermentation the wine is filtered and bottled.

In recent years the microbiology of wine making has changed. Instead of relying on spontaneous fermentation by skin-borne yeasts, many vintners now add specially selected cultures of *S. cerevisiae* to the grape juice. A number of producers now regulate the temperature of fermentation, the optimum being in the range from seven to 14 degrees C. In some areas wine is fermented not in batches but continuously. Juice is fed steadily into a fermentation process and wine is continuously removed. In general only the cheaper wines are made by continuous fermentation.

Similar methods serve for making wines from other fruit juices. The manufacture of sake, or rice wine, is more akin to making beer in that the rice contains starch rather than sugars. The starch has to be converted into fermentable sugars by means of the mold *Aspergillus oryzae.* Spores of the mold are mixed with steamed rice and the mixture is incubated for five or six days at a temperature of about 35 degrees C. to yield the product known as *koji.* Portions of *koji* are mixed with more steamed rice and some sake yeast, which is a strain of *S. cerevisiae.* This starter culture, which is called *moto,* serves to ferment the main batch of steamed rice (*moromi*) for as long as three weeks. Sake contains as much as 20 percent alcohol by volume.

The addition of brandy to wine to fortify it was originally done to arrest the yeast-fermentation process and to make the wine biologically stable. Because fortified wine contains from 15 to 20 percent alcohol by volume it is not susceptible to microbial contamination. Except for fino and amontillado sherry, fortification simply involves adding an appropriate amount of brandy to the wine, storing the wine briefly and then making a final adjustment of the alcohol content, again with brandy. In the production of fino and amontillado

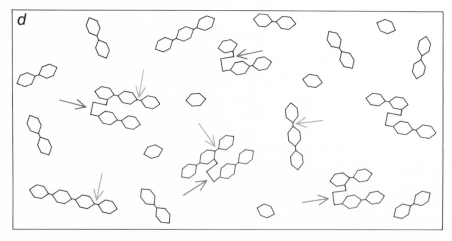

BREAKDOWN OF STARCH in brewery fermentations entails the action of the malt enzymes alpha-amylase (*black arrows*) and beta-amylase (*dark colored arrows*) on long chains of glucose, one molecule of which is shown at *a*. The straight chains (*b*) represent one of the components of starch, amylose, and the branched structures represent the other component, amylopectin. Beta-amylase splits off two glucose units at a time (*c*) to yield maltose, a disaccharide. At the same time alpha-amylase acts deeper inside the chains to split off larger sections, which in turn are acted on by beta-amylase. The products of the action of the two amylases appear in *d*. Also shown (*c, d*) are the sites of action of the two enzymes that act on dextrin. A debranching enzyme (*gray arrows*) breaks branch linkages; amyloglucosidase (*light colored arrows*) splits off single residues of glucose from dextrins. The genetic manipulation of yeast cells has improved their ability to ferment dextrins, thereby using up more of the carbohydrates in the brew and giving rise to "light" beer, which is distinguished by being low in carbohydrates.

sherry in the Jerez district of Spain the wine after fortification is matured in contact with the air to encourage the growth of a surface flora made up of a variety of yeasts. The metabolic activity of these yeasts contributes to the characteristic nutty flavor of the sherries.

The making of grain-based distilled spirits differs from the making of beer (apart from the distillation step itself) chiefly in the absence of a boiling stage. Therefore the enzymes that are active in the mash continue to operate during fermentation, breaking down more sugar and producing more alcohol. Distilled spirits differ from one another in the nature of the distillation process. Scotch malt whisky is distilled in small pot stills, whereas most other whiskeys are distilled in plants that operate continuously. With many spirits the fermented liquid is transferred to the still along with the yeast, since it has been shown that the yeast can contribute to the array of flavor compounds in the final distilled beverage. Also contributing to the flavor of the final distilled product are compounds extracted by the liquid from the wood barrels in which such spirits as whiskey and brandy are aged for a period of years.

I noted above that one of the main advantages of encouraging microbial growth in foods of the tempeh type is the increase in the protein content of the food. A logical extension of this idea is to grow suitable microorganisms on a large scale as a direct source of human food and animal feed. This was first done as a result of the shortage of food in Germany during World War I. In Berlin, Max Delbrück (not the late molecular biologist) and his colleagues developed processes for growing brewer's yeast (*S. cerevisiae*) on a large scale. Such production managed to replace as much as 60 percent of the foodstuffs Germany had been importing before the war. The yeast was incorporated mainly into soups and sausages.

Food yeast again made an important contribution to the diet in Germany in World War II. Special strains of food yeast (*Candida arborea* and *C. utilis*) were made in several production centers. During the 1960's the concept attracted a good deal of interest as a means of relieving food shortages in underdeveloped countries. Several large oil companies worked out processes for growing strains of *Candida lipolytica* in which the carbon and the energy for growth were provided by the alkanes (straight-chain hydrocarbon molecules) of petroleum. *C. lipolytica* resembles the food yeast *C. utilis*, but it has the additional property of being able to grow on alkanes.

It was at about this time that the term single-cell protein was coined to describe the new range of microbial food

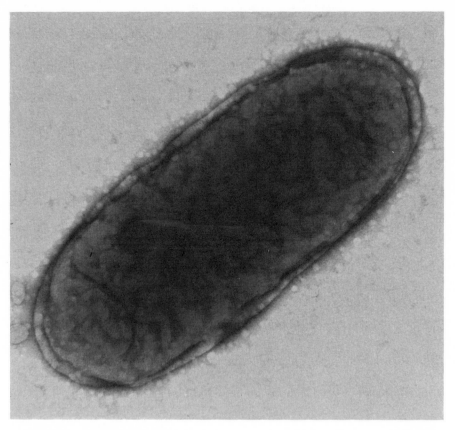

MICROBIAL FOOD AND FEED named Pruteen is based on the large-scale cultivation of the bacterium *Methylophilus methylotrophus* by Imperial Chemical Industries in Britain. A single specimen appears in this electron micrograph; the enlargement is 78,000 diameters.

SAMPLE OF PRUTEEN is in pellet form; the single-cell-protein food and feed is also made in granular form. Its color is light brown. The material is made by growing large numbers of *M. methylotrophus* bacteria, which are able to employ methane as a source of energy and carbon. In the industrial process methane is converted into methanol before it is fed to the bacteria.

and feed. The main pioneering work was done by the British Petroleum Company. The product was named Toprina, and the process was developed to the point where a $100-million plant to make it was built in Sardinia. Unfortunately the plant will never make the expected contribution to the relief of food shortages. The rising cost of petroleum was a factor (it forced many other petroleum companies to withdraw from ventures in single-cell protein), but political problems intervened too, and the company was unable to convince the Italian authorities that Toprina was toxicologically safe. The plant remains idle.

Nevertheless, at least two companies retain interest in single-cell-protein projects based on methane as the substrate. They are Hoechst AG in West Germany and Imperial Chemical Industries in Britain. The British company has recently started production of the bacterium *Methylophilus methylotrophus* in a plant capable of turning out 75,000 tons per year. The bacterium can oxidize methane, but because of safety problems that can arise with mixtures of methane and air, methane is converted chemically into methanol, which then serves as the source of carbon and energy for the bacteria. The product is named Pruteen.

The rise and fall of the single-cell-protein venture largely reflects the impact of market forces. The rising cost of petroleum has made the product less competitive with its main rivals as a cheap source of animal protein: soybeans and fish meal. In the U.S.S.R., where comparable market forces do not operate, 86 plants making single-cell protein are reported to be operating; at least 12 of them are said to rely on hydrocarbons as the source of carbon and energy for the growing cells.

For nearly a century the microbiologists working with food and beverages have been trying to understand more fully the role of microorganisms in fermentation. With some foods, such as the tempehs, little is known so far beyond the names of the fermentative organisms. With others the understanding of the changes brought about by microbial activity is much deeper. In these instances a search has usually been made for strains of microorganisms that can achieve the desired changes in the food or beverage more efficiently.

Improved strains have been selected empirically, of course, for centuries, but modern knowledge enables the food-and-beverage microbiologist to look for strains that produce, for example, certain flavoring compounds in higher or lower concentrations. Until recently the relatively sophisticated techniques for selecting improved strains could be applied only to microorganisms that are amenable to genetic analysis. For example, some of the strains of *Saccharomyces cerevisiae* that go into the fermentation of foods or beverages can be induced to form sexual spores, thereby making possible a program of yeast-strain selection based on hybridization. Such programs, which are necessarily empirical, have made a valuable contribution to the improvement of strains of yeast employed in breadmaking.

Newer techniques that manipulate *S. cerevisiae* genetically have recently been applied in the brewing industry. An example is an effort to change the ability of strains to ferment carbohydrate. The carbohydrates in malt wort consist of about 53 percent maltose, 12 percent glucose, 13 percent maltotriose, 22 percent dextrins and a trace of maltotetraose. The top- and bottom-fermenting strains of brewer's yeast are able to ferment all of them except the dextrins and occasionally the maltotetraose. Because

of the widespread rise in the demand for "light" beer, meaning beer with a lower content of carbohydrates, efforts have been made to introduce into brewer's yeast the genetic information for fermenting dextrins, which make up such a substantial part of the carbohydrate content in wort. Fortunately *Saccharomyces diastaticus,* a species related to brewer's yeast, is able to ferment dextrin. The chief aim of the recent research has therefore been to introduce genes for the fermentation of dextrin into strains of *S. cerevisiae,* the standard brewer's yeast.

Several brewing companies have been successful in introducing the genes that break down dextrin into their own favored strains of yeast. So far, however, none of these genetically engineered strains has been put to work in the brewing of beer. The reason is that when genes from *S. diastaticus* are made part of brewer's yeast, the recipient strains thereafter produce beer with an un-

pleasant phenolic flavor.

It has been shown that the unpleasant flavor is caused by a compound, 4-vinyl guaiacol, that the yeast makes from a compound derived from wort. Originally it was thought the genes for making 4-vinyl guaiacol were closely linked to the genes regulating the breakdown of dextrin, so that the off flavor was unavoidable. Recent work by Roy Tubb and his associates at the Brewing Research Foundation in Britain has shown, however, that the two groups of genes do not always remain attached when strains of yeast are genetically engineered. The path to the construction of dextrin-utilizing strains of yeast for brewing beer therefore seems to be clear. It is likely to be the first of many programs in which genes are specifically tailored to achieve some desired flavor, quality or production technique in fermented foods and beverages.

5

The Microbiological Production
of Pharmaceuticals

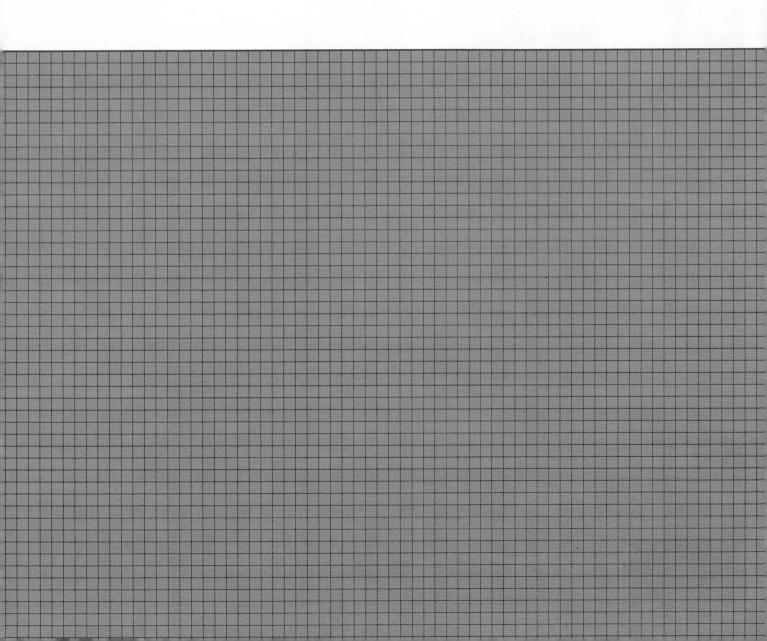

The Microbiological Production of Pharmaceuticals

BY YAIR AHARONOWITZ AND GERALD COHEN

The introduction of penicillin opened up a new era in medicine. Now microorganisms manufacture not only a host of other antibiotics but also vitamins, hormones, alkaloids, antitumor drugs and interferons

The introduction of microbiology into the pharmaceuticals industry, which began in the 1940's, has brought about a transformation profound enough to be called a revolution. Advances in our understanding of microorganisms and techniques for manipulating them genetically are now routinely exploited in the identification of new therapeutic substances, in research and development and in the processes of industrial production itself. The linked chemical reactions that make up the metabolic system of a microorganism constitute the means of production. In huge tanks cultures of genetically identical cells bred for high yield are immersed in a rich liquid medium. The pharmaceutically valuable products of metabolism are later extracted and subjected to further processing.

Such cell cultures on a mighty scale are employed in the pharmaceuticals industry in three ways, which can be differentiated on the basis of how much of the information needed to make the product is present in the microorganism's unaltered genome (complete set of genes). In the case of the antibiotics the product is a natural metabolite, and all the information for its synthesis is native to the cell. (Even so, the product is often chemically modified later.) It was the identification of penicillin, a natural metabolite of the mold *Penicillium,* that initiated the transformation of the pharmaceuticals industry.

Commercially and clinically the anti-biotics are the most important class of pharmaceuticals made by microbiological techniques. Similar techniques have also been adapted, however, to the production of substances that are not natural metabolites of microorganisms. In the manufacture of steroid hormones, for example, microorganisms carry out individual steps called bioconversions in a long sequence of synthetic processes; the other steps are accomplished by nonbiological methods. Only the information for the few biological steps resides in the genome of the organism.

In the third approach none of the information that defines the structure of the product molecule is initially found in the genome of the microorganism; the information is inserted into the cell. In this way bacterial or fungal cells can be made to produce human proteins. Methods of this kind are now being explored for the manufacture of such clinically important pharmaceuticals as insulin. Although these techniques are the newest and most glamorous in the pharmaceuticals industry, they are being assimilated into a field where microbiological methods have already brought forth a huge commercial enterprise. In 1979 the wholesale value of prescription drugs sold in the U.S. was about $7.5 billion; of this amount some 20 percent, or $1.5 billion, represented sales of drugs in whose production microorganisms played a significant role.

The largest class of pharmaceuticals consists of those in which most or all of the required genetic information is present in the unaltered genome of the cell. The antibiotics are the most important members of this class economically, but also included are viral and bacterial antigens, antifungal agents, certain antitumor drugs, alkaloids and vitamins. In 1978 the worldwide bulk sales of the four most important groups of antibiotics—the penicillins, the cephalosporins, the tetracyclines and erythromycin—amounted to $4.2 billion. (The sales are given for 1978 because that is the most recent year for which complete information is available for the international market; the amount has been adjusted to 1980 prices to compensate for the effects of inflation.) Another commercially important group of antibiotics consists of the aminoglycosides, which include streptomycin. After the antibiotics the pharmaceuticals with the next-highest sales were the vitamins; the wholesale value of the six most important vitamins in 1978 (again based on 1980 prices) was $670 million.

The industrial process that underlies this market is fermentation. The fermenters, or tanks, in which the metabolic manufacture of pharmaceuticals proceeds have a maximum volume of about 100,000 liters. Cultures of industrial strains of fungi or bacteria are started in smaller tanks, then transferred to the large fermenters, where strict control is maintained over the temperature, the *p*H, the oxygen supply and the nutrients in the culture medium. The mixture is stirred by blades inside the fermenter.

The microorganisms that carry out fermentation in the antibiotics industry are drawn from a rather narrow taxonomic range. János Berdy of the Research Institute for Pharmacological Chemistry in Budapest has classified them in three main groups. Six genera of filamentous fungi give rise to almost 1,000 distinct antibiotics. Among these fungi are molds of the genus *Cephalosporium,* which yield cephalosporins, and

MANUFACTURE OF PENICILLINS is accomplished in a combination of biological and chemical steps; shown here are crystallizers, the site of one of the key processes in production. Manufacture begins in fermentation tanks with a capacity of as much as 100,000 liters. In the tanks an industrial strain of the fungal mold *Penicillium chrysogenum* is grown in a rich liquid medium; a form of penicillin called penicillin *G* is a natural metabolite of the fungal cells. In this plant, operated by Pfizer, Inc., in Groton, Conn., as many as 15 fermentation tanks are linked on a staggered production schedule to provide a continuous output of the antibiotic. When fermentation, which takes several days, is complete, penicillin *G* is separated from the spent mold cells and injected into the crystallizers, where butanol is added. The butanol is evaporated, carrying water with it and leaving behind a crystalline slurry of penicillin *G* of more than 99 percent purity. Subsequent chemical modifications yield other forms of penicillin.

CATEGORY OF DRUG	MAJOR U.S. PRODUCERS	MARKET VALUE
PENICILLINS	Ayerst Laboratories Lederle Laboratories Eli Lilly and Company Smith, Kline & French Laboratories E. R. Squibb & Sons, Inc. Warner-Lambert Company Wyeth Laboratories	$220,943,000
OTHER BROAD- AND MEDIUM-SPECTRUM ANTIBIOTICS	Abbott Laboratories Bristol Laboratories Lederle Laboratories Eli Lilly and Company Merck Sharp & Dohme Schering-Plough Corporation E. R. Squibb & Sons, Inc. The Upjohn Company Warner-Lambert Company Wyeth Laboratories	$638,297,000
ANTIBIOTICS IN COMBINATION WITH SULFONAMIDES	Bristol Laboratories Burroughs Wellcome Co. Ross Laboratories	$16,921,000
TOPICAL ANTIBIOTICS	Lederle Laboratories Eli Lilly and Company Marion Laboratories, Inc. Schering-Plough Corporation The Upjohn Company Warner-Lambert Company	$17,064,000
VACCINES	Lederle Laboratories Merck Sharp & Dohme Warner-Lambert Company Wyeth Laboratories	$90,000,000
SULFONAMIDES	Alcon Laboratories, Inc. Lederle Laboratories Hoffmann–La Roche, Inc. Smith, Kline & French Laboratories E. R. Squibb & Sons, Inc. Warner-Lambert Company	$47,562,000
ANTIFUNGAL DRUGS	Ayerst Laboratories Barnes-Hind Pharmaceuticals, Inc. Ciba-Geigy Corporation Lederle Laboratories Ortho Pharmaceutical Corporation Hoffmann–La Roche, Inc. Schering-Plough Corporation E. R. Squibb & Sons, Inc.	$103,911,000
ANTISEPTIC PREPARATIONS	Burroughs Wellcome Co. Norwich-Eaton Pharmaceuticals Ortho Pharmaceutical Corporation Sterling Drug Inc. E. R. Squibb & Sons, Inc.	$15,000,000
TUBERCULOSTATIC AGENTS	Ciba-Geigy Corporation Dow Chemical U.S.A. Lederle Laboratories E. R. Squibb & Sons, Inc. Warner-Lambert Company	$12,835,000
DIGESTIVE ENZYMES	Armour and Company B. F. Ascher & Company, Inc. Hoechst-Roussel Pharmaceuticals, Inc. Organon Inc. Reed & Carnrick Warner-Lambert Company	$16,999,000
VITAMINS (PRESCRIPTION ONLY)	Abbott Laboratories The Central Pharmacal Company Lederle Laboratories Mead Johnson & Company Hoffmann–La Roche, Inc. Ross Laboratories E. R. Squibb & Sons, Inc. Warner-Lambert Company	$133,891,000

SALES OF THREE CATEGORIES OF PHARMACEUTICALS in whose production microorganisms play a significant role—anti-infective agents, enzymes and vitamins—are dominated by the systemic antibiotics. The data give the wholesale value of pharmaceuticals sold in the U.S. by American companies in 1979. Of the anti-infective agents some are produced microbiologically and some are not. The antibiotics are manufactured almost entirely by fermentation. Those of commercial importance other than penicillin include the cephalosporins, tetracyclines, erythromycin and streptomycin. The data for vaccines and antiseptic preparations are estimates. Before the introduction of antibiotics the sulfonamides were the main anti-infective drugs available to the physician. They now hold a small place in the market. Of the prescription vitamins some are made by fermentation and some by nonbiological methods.

the genus *Penicillium*, source of the penicillins. Among the nonfilamentous bacteria two genera synthesize roughly 500 antibiotics. By far the largest number of antibiotic substances come from the actinomycetes, a group of filamentous bacteria. Three genera of actinomycetes account for almost 3,000 antibiotic agents. Actinomycetes of the genus *Streptomyces* make the largest proportion of them, including the tetracyclines.

The number of antibiotic substances made by each genus does not bear much relation to clinical or commercial importance. Of the almost 5,000 antibiotics known only about 100 have been marketed. The majority of these are derived from the streptomycetes, which as of 1977 yielded 69 products. It is the penicillins and the cephalosporins, however, that dominate commerce in antibiotics. Of the $4.2 billion in world bulk sales of antibiotics in 1978 about $1 billion is attributed to sales of penicillins and $.5 billion to sales of cephalosporins. Most of the remaining sales were accounted for by products of the actinomycetes, including about $1 billion worth of tetracyclines. No bacterial product had a substantial share of the market, although some bacterial antibiotics are useful in particular clinical situations.

Although the taxonomic range of the organisms that make antibiotics is narrow, the molecules themselves are extremely diverse both in chemical structure and in physiological function. Antibiotics have been identified that interfere with almost every phase of the life cycle of a bacterial cell; a few have been found that attack fungal cells. The penicillins and cephalosporins interfere with the assembly of the bacterial cell wall; the polyene macrolides, such as amphotericin *B*, disturb fungal membrane functions; the bleomycins and anthracyclines interfere with DNA replication; the rifamycins interrupt the transcription of DNA into messenger RNA; erythromycin, the tetracyclines and streptomycin disable the ribosomal complex (the site of protein synthesis).

Because of the diverse structures and functions of the antibiotics it is not easy to define them; a practical definition of an antibiotic is a microbial product of low molecular weight that specifically interferes with the growth of microorganisms when it is present in exceedingly small amounts. Most of the substances that satisfy this definition are fungal or bacterial metabolites, which have no obvious role in the growth and maintenance of the cell. These molecules are called secondary metabolites, to distinguish them from the primary metabolites needed in the growth of the organism. Alkaloids, toxins and pigments are also secondary metabolites. Such compounds are formed only after the growth of the cell has slowed and the

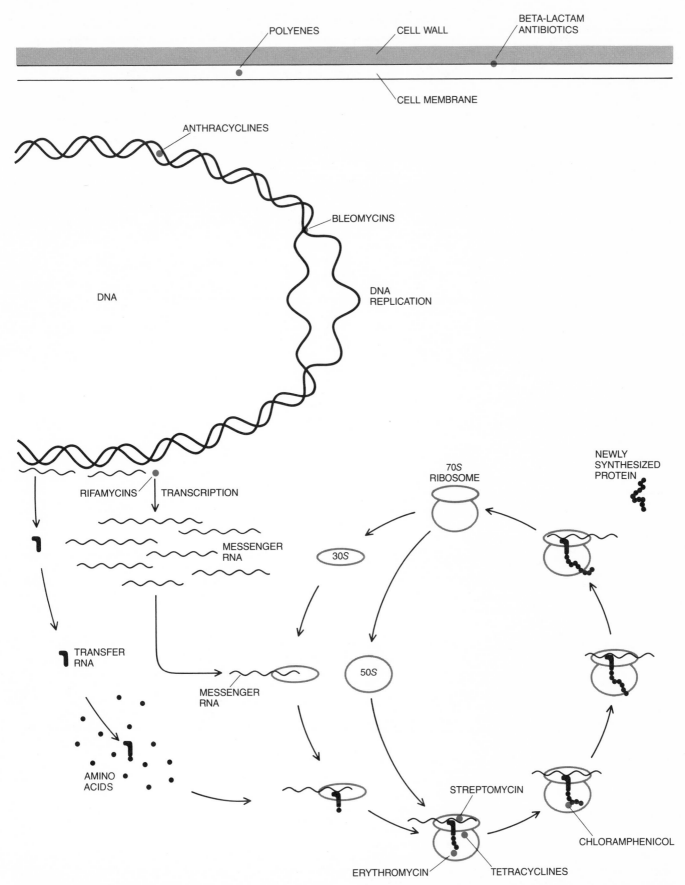

POLYENES

CELL WALL

BETA-LACTAM ANTIBIOTICS

CELL MEMBRANE

ANTHRACYCLINES

BLEOMYCINS

DNA

DNA REPLICATION

70S RIBOSOME

NEWLY SYNTHESIZED PROTEIN

RIFAMYCINS   TRANSCRIPTION

MESSENGER RNA

30S

50S

TRANSFER RNA

MESSENGER RNA

AMINO ACIDS

STREPTOMYCIN

CHLORAMPHENICOL

ERYTHROMYCIN   TETRACYCLINES

SITES OF ACTION of the antibiotics are extremely diverse, including almost every important process in the life of a bacterial cell. The penicillins and the cephalosporins interrupt the construction of the bacterial cell wall. The bleomycins and the anthracyclines interfere with the replication of DNA. The rifamycins prevent DNA from being transcribed into messenger RNA. Erythromycin, streptomycin, chloramphenicol and the tetracyclines all disable the ribosomal complex, where messenger RNA is translated into protein. These meta- bolic processes are subtly different in bacteria and in mammalian cells, and so antibiotics are toxic for microorganisms but safe for human beings. The treatment of fungal infections and of cancer is currently less effective than that of bacterial infections because few substances with selective toxicity for tumor cells and for fungi have been found. One group of antifungal agents that has been found is made up of the substances called polyenes. These drugs, which include amphotericin *B,* interfere with the function of the fungal cell membrane.

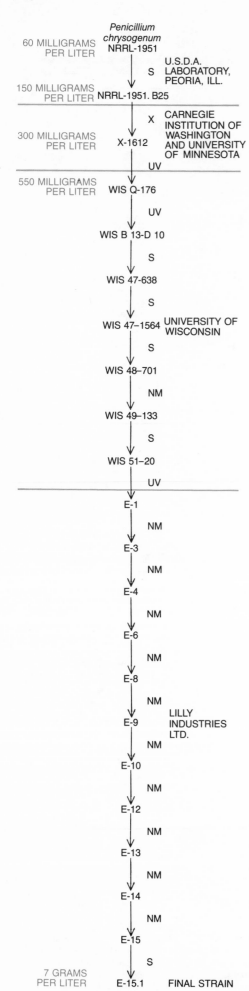

60 MILLIGRAMS PER LITER	*Penicillium chrysogenum* NRRL-1951
	S — U.S.D.A. LABORATORY, PEORIA, ILL.
150 MILLIGRAMS PER LITER	NRRL-1951. B25
	X — CARNEGIE INSTITUTION OF WASHINGTON AND UNIVERSITY OF MINNESOTA
300 MILLIGRAMS PER LITER	X-1612
	UV
550 MILLIGRAMS PER LITER	WIS Q-176
	UV
	WIS B 13-D 10
	S
	WIS 47-638
	S
	WIS 47-1564 — UNIVERSITY OF WISCONSIN
	S
	WIS 48-701
	NM
	WIS 49-133
	S
	WIS 51-20
	UV
	E-1
	NM
	E-3
	NM
	E-4
	NM
	E-6
	NM
	E-8
	NM
	E-9 — LILLY INDUSTRIES LTD.
	NM
	E-10
	NM
	E-12
	NM
	E-13
	NM
	E-14
	NM
	E-15
	S
7 GRAMS PER LITER	E-15.1 — FINAL STRAIN

cell has entered the stage of its life cycle called the idiophase. The function of antibiotics in the cells that make them is not clear, although it has been suggested they serve to inhibit the growth of competing microorganisms.

Like other secondary metabolites, an antibiotic is the end product of a long series of enzymatically catalyzed reactions. Many genes, both structural and regulatory, contribute to the synthesis; molecular precursors must also be synthesized. The complexity of the metabolic pathways leading to the manufacture of an antibiotic has important consequences for industrial production and for research methods employed to improve the commercial product.

Although antibiotics have a wide range of chemical structures and varied sites of action, they all satisfy the principle of selective toxicity formulated early in this century by Paul Ehrlich. The principle holds that an effective chemotherapeutic agent should be safe for human tissues but toxic to the infecting organism. Although the fundamental processes of cell metabolism in the human body are similar to those in much simpler organisms, subtle differences can make an antibiotic lethal to the infecting agent but harmless to the patient. This discrimination is an essential property of antibiotic action. Many substances have been found to exhibit selective toxicity for bacteria, but there has been conspicuously less success in the search for agents effective against fungi, viruses, parasites or tumor cells.

Penicillin's toxicity for bacteria was first noted by Alexander Fleming in the 1920's; that the effect is selective was demonstrated by Howard W. Florey, Ernst B. Chain and their colleagues at the University of Oxford in 1941, when they showed that penicillin could cure bacterial infections. Penicillin is harmless in man because the site at which it acts—the bacterial cell wall—has no exact equivalent in a human cell. Neither

REPEATED MUTATIONS were necessary to create a strain of the mold *Penicillium chrysogenum* that synthesized enough penicillin to form the basis of a commercial process. Radiation and chemical agents were employed by four groups of investigators to induce mutations in the mold. ("S" stands for spontaneous mutation, "X" for X-radiation, "UV" for ultraviolet radiation and "NM" for nitrogen mustard.) Selection of the superior strains ultimately gave rise to strain *E* 15.1, which yielded 55 times as much penicillin as laboratory strains. Simultaneous improvements in fermentation technique increased yields still further; yield figures in this chart reflect both kinds of increase. Classical genetic techniques such as these are still important in the antibiotics industry; complexity of antibiotic synthesis in microorganisms makes it impractical to develop new strains by directly altering single genes. Current fermentation methods yield more than 20 grams per liter.

of these findings by themselves, however, would have led to a practical antibiotic for clinical purposes; the amount of penicillin made by laboratory strains of *Penicillium* molds, a few milligrams per liter of culture, was far too small to form the basis of an industrial process.

In an attempt to improve the yield a highly productive strain of *Penicillium chrysogenum* was exposed systematically to a variety of mutagens, including nitrogen mustard, ultraviolet radiation and X-radiation; advantage was also taken of spontaneous mutations. After each round of exposure the next generation of mold cells was examined for mutants with higher productivity. The process was repeated at length; 21 rounds of mutation and selection carried out in a number of laboratories were needed to increase the yield of penicillin by a factor of 55. Combined with improvements in fermentation technique this was sufficient for the first commercial production. Since then further selection and improvements in fermentation technology have raised the efficiency of manufacture to 20 grams per liter or more, an improvement of 10,000-fold over the yield in Florey's laboratory.

Classical genetic techniques that rely on random mutation are cumbersome and time-consuming, mainly because mutations that increase antibiotic yields offer no advantage to the microorganism. It is therefore necessary to screen many colonies of survivors to determine their yields under fermentation conditions. In addition productive mutants appear infrequently. Although more sophisticated methods aimed at altering single genes are now available, classical methods are still indispensable for improving antibiotic yields.

The reason for the retention of older methods lies in the nature of secondary-metabolite synthesis. Unlike a protein, which is the immediate product of a single gene, an antibiotic is made by the joint action of the products of between 10 and 30 genes. For most commercial antibiotics the entire pathway has not been worked out. As a result attempts to alter single genes are for the most part not effective in increasing yields. The one major recent modification of genetic screening has been the development of automatic methods for examining the survivors of the process of inducing mutations and identifying new strains with higher yields. The automatic equipment screens tens of thousands of survivor types per round of mutation.

Penicillin is dramatically effective against a wide range of Gram-positive bacteria. (Gram-positive and Gram-negative bacteria are distinguished on the basis of a staining procedure developed in 1884 by Hans Christian Joachim Gram; the test is sensitive to fundamental differences in the cell walls of

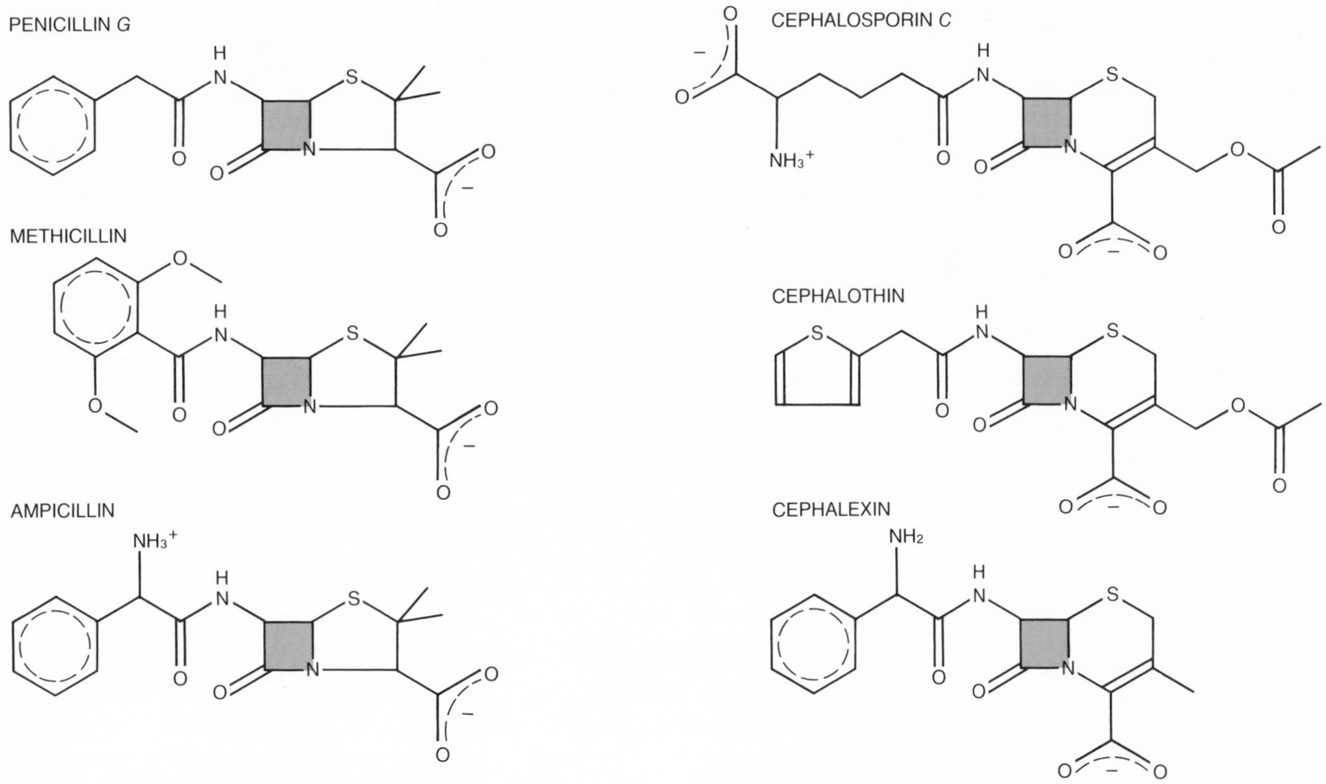

PENICILLIN G

METHICILLIN

AMPICILLIN

NH_3^+

CEPHALOSPORIN C

NH_3^+

CEPHALOTHIN

CEPHALEXIN

NH_2

CHEMICAL STRUCTURE AND FUNCTION of the penicillins and the cephalosporins hinge on the four-member beta-lactam ring (*color*); the drugs are called beta-lactam antibiotics. The ring is essential to the action of these compounds in halting the construction of the bacterial cell wall. Side groups attached to the ring can increase the potency of the antibiotic and improve its pharmacological properties. In the penicillins a single side group varies. In commercial production penicillin G serves as a core structure for the attachment of new side chains after the removal of the benzyl group. Methicillin is resistant to inactivation by bacterial enzymes; ampicillin is effective against Gram-negative bacteria. Both of these improvements over penicillin G are accomplished by the alteration of the side group. The cephalosporins possess two variable side chains. Cephalosporin C is employed as a core structure in a way analogous to penicillin G.

the bacteria.) When penicillin was introduced into clinical practice, it was found that many common bacterial infections, such as streptococcal pharyngitis, pneumococcal pneumonia and most staphylococcal infections, could be cured rapidly and completely. Penicillin also cured serious and frequently fatal infections such as meningococcal meningitis, and it was effective in treating some forms of bacterial endocarditis that had invariably been fatal.

These striking clinical results stimulated a search for additional naturally occurring antibiotics. The search was motivated by two factors. Penicillin was much less effective against the Gram-negative bacteria. It was also observed that certain Gram-positive bacteria possessed enzymes capable of inactivating penicillin; thus the bacteria were resistant to the antibiotic. By means of a newly developed technique for screening the microorganisms present in soil, Selman A. Waksman and his colleagues at Rutgers University isolated streptomycin and other antibiotics from ac-

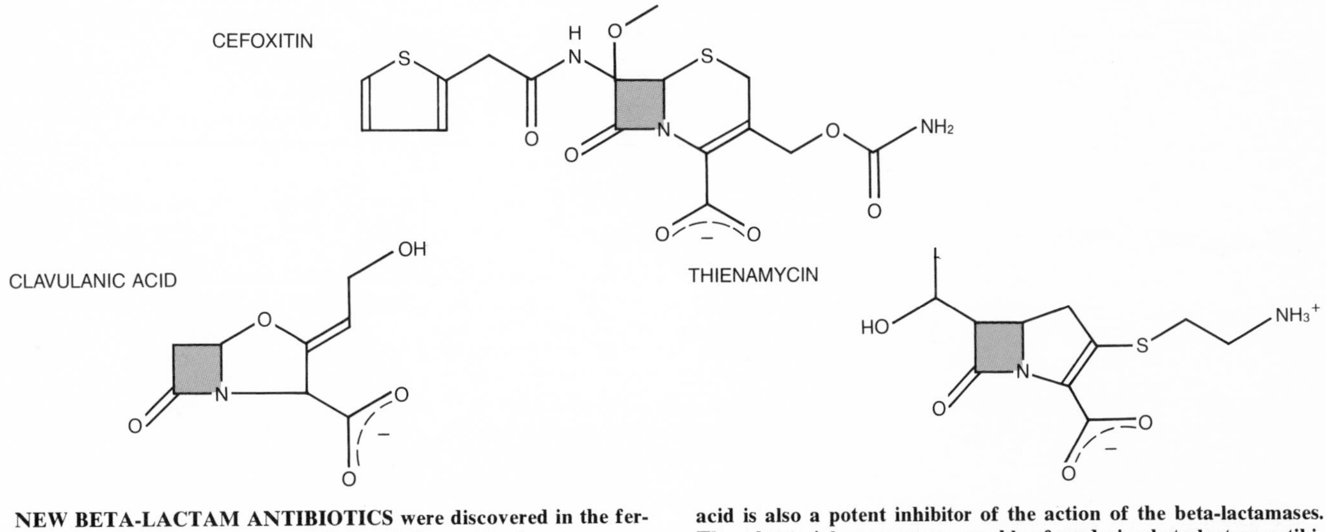

CEFOXITIN

NH_2

CLAVULANIC ACID

OH

THIENAMYCIN

HO

NH_3^+

NEW BETA-LACTAM ANTIBIOTICS were discovered in the fermentation broths of microorganisms of the genus *Streptomyces*, a subgroup of the filamentous bacteria known as actinomycetes. Until the streptomycete products were isolated the fungal molds *Penicillium* and *Cephalosporium* had been the only sources of beta-lactam antibiotics. All three molecules possess antibiotic activity; clavulanic acid is also a potent inhibitor of the action of the beta-lactamases. These bacterial enzymes are capable of rendering beta-lactam antibiotics ineffective by splitting open the beta-lactam ring. Clavulanic acid is now being marketed in combination with the beta-lactam drug amoxicillin; this pharmaceutical hybrid, known as augmentin, is a potent antibiotic that is also resistant to beta-lactamase inactivation.

tinomycetes of the genus *Streptomyces;* some of these preparations were effective against Gram-negative bacteria, and others against Gram-positive bacteria. The mold *Cephalosporium acremonium* was isolated in 1945 from the sea off Sardinia by G. Brotzu of the Institute of Hygiene of Cagliari. The cells of the mold were found to synthesize several related antibiotics, one of which, named cephalosporin C, was particularly effective against penicillin-resistant Gram-positive pathogens.

Although the penicillins, the cephalosporins and streptomycin were the most important discoveries of the early period of antibiotic identification, there were many others. The number of new antibiotics identified each year increased in a roughly linear way from the late 1940's through the early 1970's, when about 200 new substances per year were being characterized. By the end of the 1970's new antibiotics were being found at a rate of about 300 per year, of which roughly 150 were products of the actinomycetes.

The proportion of the discoveries that were put into commercial production, however, declined rapidly after the 1950's. It became increasingly difficult to isolate a new antibiotic sufficiently superior to an existing product to warrant its introduction into clinical practice. As a result of this diminishing return and the development of resistance to antibiotics in many bacteria the focus of research shifted. By the mid-1960's most work was directed toward modifying the structure of existing antibiotics to increase their potency, protect them from bacterial inactivation and improve their pharmacological properties.

Most of the effort was focused on the penicillins and the closely related cephalosporins. In both groups the central structure of each molecule is the four-member beta-lactam ring, composed of three carbon atoms and a nitrogen atom; the penicillins and the cephalosporins are collectively known as the beta-lactam antibiotics. In addition to having a broad spectrum of antibacterial activity the beta-lactam antibiotics are probably the least toxic of all the major groups of antibiotics.

Although the exact mechanism by which the beta-lactam antibiotics destroy bacteria has not been completely elucidated, it is clear that they interrupt the manufacture of the cell wall. Specifically they interfere with both the synthesis and the assembly of peptidoglycan, the major constituent of the wall. They do so by attaching themselves to at least three enzymes—transpeptidase, carboxypeptidase and endopeptidase—that catalyze the polymerization and insertion of peptidoglycan into the wall. Disruption of this process leads rapidly to the dissolution and death of the cell. The ubiquity of peptidoglycan in the cell wall of prokaryotes and its absence in higher organisms are responsible for the highly selective toxicity of the beta-lactam antibiotics.

For more than 30 years two molds, *Penicillium chrysogenum* and *Cephalosporium acremonium,* were the exclusive sources of the beta-lactam antibiotics. Recently, however, in an intensive screening program of prokaryotic soil organisms undertaken by Eli Lilly and Company and Merck, Sharp & Dohme, new beta-lactam antibiotics were found in the fermentation broths of streptomycetes. These compounds, the cephamycins, have a structure similar to that of the cephalosporins, with the addition of a methoxyl group (CH_3O-) on the beta-lactam ring. In some instances the side group increases effectiveness against both Gram-negative and penicillin-resistant organisms.

In the 1960's and 1970's efforts to improve the beta-lactam antibiotics focused on adding new side groups to the beta-lactam ring. The semisynthetic approach, which is now widely adopted in manufacturing penicillins and cephalosporins, relies on chemical synthesis to substitute one side chain for another after fermentation has produced a molecule with the central ring. The addition of new side chains, an approach that has also been taken with the aminoglycosides, including streptomycin, can improve the potency, lack of toxicity and stability of the substance; it can also broaden the spectrum of organisms against which the antibiotic is effective.

In the semisynthetic manufacture of penicillin an industrial strain of *Penicillium chrysogenum* is grown in the presence of phenylacetic acid, which results in the formation of penicillin G. The production of penicillin G is carried out on a large scale. A number of fermentation tanks are usually operated on a staggered schedule to provide a virtually continuous yield for the recovery and modification processes. The plant operated by the Dutch company Gist-Brocades NV in Delft, for example, has 14 fermenters, each with a capacity of 100,000 liters. The time required for fermentation is 200 hours; the recovery process takes 15 hours, and so the 14 fermentation tanks allow the re-

CROSS-LINKAGE OF PEPTIDOGLYCAN CHAINS in the formation of the bacterial cell wall is interrupted by the beta-lactam antibiotics. Each chain consists in part of alternating units of the amino sugars N-acetylglucosamine (NAG) and N-acetylmuramic acid (NAM). The NAM units are attached to polypeptide groups. The cross-linking of the polypeptides to one another by peptide bonds gives the cell wall its rigidity. In the bacterium *Staphylococcus aureus* the linkage is accomplished when the final glycine unit of one chain is inserted into the bond linking two alanine units of another chain. The final alanine unit is cleaved away and a new bond is formed between the next alanine and the final glycine. The creation of this bond is catalyzed by enzymes called transpeptidases and carboxypeptidases. By binding to the peptidases and inactivating them the beta-lactam antibiotics prevent the linking of the peptidoglycan chains. These antibiotics are thus damaging to growing cells that are forming cell walls.

PENICILLIN G

6-AMINOPENICILLANIC ACID

PENICILLIN ACYLASE →

$(CH_3)_2SiCl_2 + PCl_5$ −40° C.

H_2O 0° C.

BUTANOL
−40° C.

Si $(CH_3)_2$

MICROBIOLOGICAL CONVERSION of penicillin *G* into 6-aminopenicillanic acid (6-APA) has fewer steps and is cheaper than the chemical conversion. In the production of semisynthetic penicillins 6-APA forms a chemical nucleus to which side chains can be attached, yielding new antibiotics. In performing the chemical conver-sion three steps must be carried out at low temperature and under strictly anhydrous conditions with a number of chemical solvents. The biological process relies on bacteria that make enzymes called acylases; the enzymes cleave away the benzyl group, leaving 6-APA. The fermentation can be carried out in water at 37 degrees Celsius.

covery to proceed without interruption.

The fermentation is a "fed batch" process, in which a sugar solution is added to the fermentation broth continuously. Phenylacetic acid is the precursor of the benzyl side chain of penicillin *G*. When the fermentation is complete, the thickened broth is passed through a rotating filter to separate the mold cells from the liquid medium that contains the penicillin; the cells are then washed. The filtrate and the washings are put through a chemical extractor. A source of potassium ions is added to the mixture, and the result is the crystalline potassium salt of penicillin *G*. The filtered and dried salt constitutes a stream of the antibiotic of 99.5 percent purity.

Following recovery the penicillin *G* is injected into the fermentation broth of a strain of bacteria that secrete the enzymes called acylases. The acylases selectively remove the benzyl group from the molecule, yielding 6-aminopenicil-lanic acid, or 6-APA. This core structure has weak antibacterial properties. It also forms a convenient molecular center for the attachment of the side groups that are capable of increasing the potency of the antibiotic.

The semisynthetic approach has also been adopted in the manufacture of the cephalosporins. The starting point is cephalosporin *C,* made by *Cephalosporium acremonium.* Cephalosporin *C* acts against both Gram-negative and Gram-positive bacteria, but the activity is too weak for clinical purposes. When an amide side chain of the molecule is removed, however, the remaining molecular core, 7-alpha-aminocephalosporanic acid, or 7-ACA, is useful in the way 6-APA is. By the attachment of appropriate side chains a variety of semisynthetic cephalosporins are created.

Effective as the beta-lactam antibiotics are, they can be thwarted by bacterial resistance. The biochemical basis of the resistance is the hydrolysis, or cleavage, of the bond in the beta-lactam ring called the cyclic amide bond. The hydrolysis renders the molecule incapable of attachment to the bacterial peptidases. This destruction of the antibiotic is catalyzed by the enzymes called beta-lactamases, which are widespread in bacteria, actinomycetes, cyanobacteria ("blue-green algae") and yeasts. Moreover, the enzymes can be passed from one microorganism to another by plasmids bearing the gene for them. The widespread use of antibiotics has created selective pressure favoring the survival of those microorganisms that have acquired the gene.

A critical task in antibiotic development is overcoming this enzyme-based inactivation of the beta-lactam antibiotics. At least two strategies have been pursued. One is the identification of natural antibiotics that are not susceptible to inactivation by bacterial beta-lac-

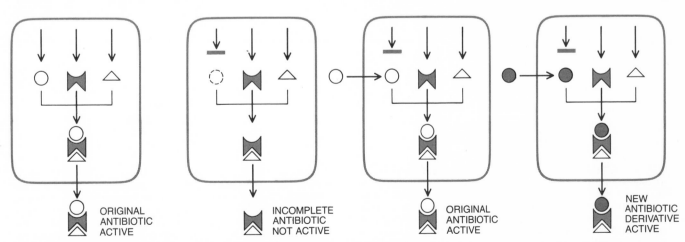

MUTATIONAL BIOSYNTHESIS, or mutasynthesis, is an elegant method of developing new antibiotics. A mutation induced in the gene that codes for one precursor of a microorganism's natural antibiotic leads to the synthesis of an incomplete antibiotic molecule, lacking a single chemical constituent. When the missing precursor is added to the medium in which the microorganism is grown, the natural antibiotic is produced. When precursors with slightly different structures are added to the medium, new antibiotics are synthesized.

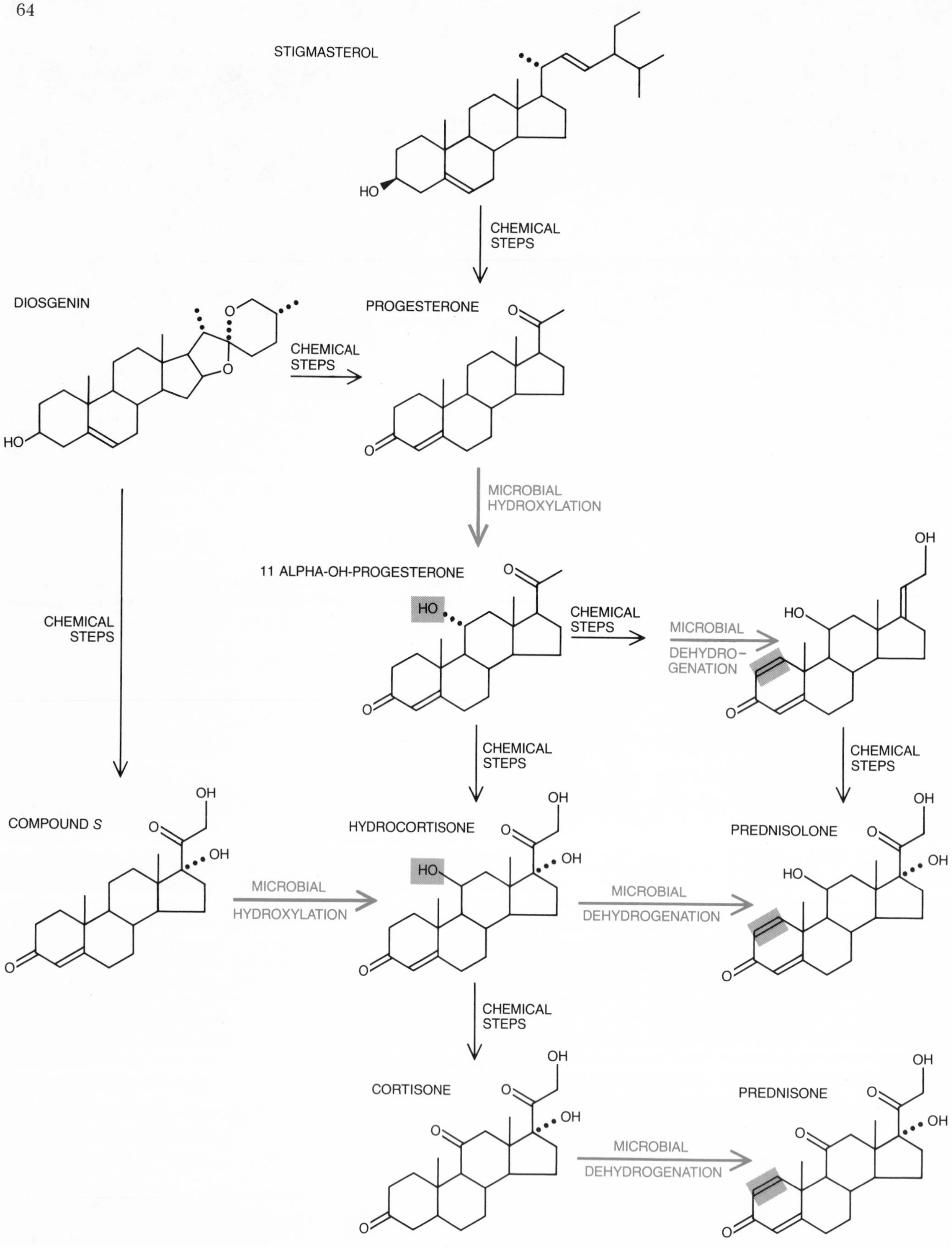

STIGMASTEROL

DIOSGENIN

CHEMICAL STEPS

PROGESTERONE

CHEMICAL STEPS

MICROBIAL HYDROXYLATION

11 ALPHA-OH-PROGESTERONE

CHEMICAL STEPS

MICROBIAL DEHYDRO–GENATION

CHEMICAL STEPS

CHEMICAL STEPS

COMPOUND S

HYDROCORTISONE

MICROBIAL HYDROXYLATION

PREDNISOLONE

MICROBIAL DEHYDROGENATION

CHEMICAL STEPS

CORTISONE

PREDNISONE

MICROBIAL DEHYDROGENATION

MANUFACTURE OF STEROIDS includes several important microbiological steps in a process made up primarily of chemical syntheses. The raw materials of production are the complex alcohols called sterols. Stigmasterol is a by-product of the soybean-oil industry; diosgenin is extracted from the roots of the Mexican barbasco plant. The sterols are converted in a series of chemical steps into one of two intermediates: compound S and progesterone. A fungal mold (either *Rhizopus nigricans* or *Curvularia lunata*, depending on the intermediate compound) is then used to hydroxylate the molecule, adding a hydroxyl group (–OH) to the four-ring steroid nucleus. This microbial step was critical in the development of a commercial-

ly feasible method of manufacturing steroids. The chemical method of hydroxylation is complex and difficult; finding a biological method reduced the complete synthesis from 37 to 11 steps and greatly reduced the unit cost of manufacture. Steroids, widely used for the control of inflammation, were thus brought within economic reach of most patients. The other important microbial step is dehydrogenation: the removal of two hydrogen atoms from the steroid nucleus. Processes resembling this generalized scheme are employed in producing cortisone, hydrocortisone, prednisolone and prednisone, which are all medically important synthetic molecules having pharmacological properties different from those of the natural steroids.

tamases. Five years ago, in the course of a search for inhibitors of peptidoglycan synthesis, workers at Merck, Sharp & Dohme discovered a new class of beta-lactam antibiotics. These compounds, the thienamycins, were found in the fermentation broths of *Streptomyces cattleya*. They are extremely potent against many Gram-positive and Gram-negative bacteria; more important, they are capable of inhibiting the beta-lactamases, thereby protecting themselves against inactivation.

The second strategy is to find substances that possess little or no antibiotic activity of their own but are potent inhibitors of the beta-lactamases. Such an inhibitor is then combined with a beta-lactam antibiotic to create a composite that resists inactivation. The beta-lactamase inhibitors clavulanic acid and the olivanic acids were identified by workers at the Beecham Pharmaceutical Company in Britain. These substances share the beta-lactam structure, but their greatest effect is the inhibition of enzymes. Recent studies by Jeremy Knowles and his colleagues at Harvard University have shown that clavulanic acid and olivanic acid inactivate the enzyme by the "suicide" method. They bind to the beta-lactamase molecule, initiating a catalytic reaction. In contrast to other beta-lactam substances, however, which are hydrolyzed and then released, the inhibitors remain jammed in the active site of the enzyme, precluding further activity. Clavulanic acid is an effective inhibitor of the beta-lactamases of most clinically important microorganisms. The new antibiotic augmentin, made by Beecham, is a combination of the beta-lactam antibiotic amoxicillin and clavulanic acid.

Along with classical genetic techniques, identification of natural metabolites and semisynthesis, several new genetic methods have recently been added to the microbiologist's armamentarium. One elegant process is mutational biosynthesis, or mutasynthesis, which consists in the creation by a specific mutation of a microorganism that is unable to synthesize one molecular precursor of an antibiotic. When that precursor is added to the cell culture, the organism's natural antibiotic is made. If another precursor with a slightly different chemical structure is added, a new antibiotic may be produced.

A second method derives from the fact that secondary metabolism is crucially dependent on the levels of primary metabolites present in the cell. Among the substrates of antibiotic synthesis are carbohydrates, amino acids, purines, pyrimidines, fatty acids and activated acetyl and propinyl molecules. The rate of production of the antibiotic may be limited by the rate of synthesis or availability of a primary metabolite. When that is the case, and the gene for

the primary metabolite has been identified, it is possible by genetic manipulation to create a strain that makes more of the precursor. The result is an increase in the yield of the antibiotic. One of us (Aharonowitz) has employed this approach to increase the yield of cephalosporins from *Streptomyces clavaligerus*.

Antibiotics are capable of inhibiting the growth of the cells that make them as well as the growth of pathogenic bacteria; normally the antibiotic is elaborated only after growth has ceased. Certain mutant strains, however, are resistant to their own antibiotics and are highly productive. For example, certain strains of *Streptomyces aureofaciens* that resist high concentrations of their own antibiotic, chlortetracycline, have turned out to manufacture it very efficiently.

Whereas the identification, modification and production of substances effective against bacterial infections has been remarkably successful, similar agents for viral and fungal infections and for tumors are still quite primitive. The drawback of most available antifungal and antitumor agents is a lack of selectivity; most are harmful to normal mammalian cells as well as to pathogenic organisms or cancerous cells. The majority of antifungal agents are toxic when they are taken internally and can only be applied topically. Several microbial metabolites inhibit tumor growth, but they too are toxic.

One antitumor substance that has merited serious attention is the glycopeptide bleomycin, isolated by Hamao Umezawa and his colleagues at the Institute of Microbial Chemistry in Tokyo from the culture broths of *Streptomyces verticillus*. It apparently acts by binding to and breaking the DNA of the tumor cells and interfering with the replication of both DNA and RNA. Another clinically important group of antitumor drugs is composed of an aminoglycoside unit and an anthracycline molecule. These substances also inhibit DNA and RNA synthesis. Both bleomycin and the anthracyclines, however, are potentially damaging to the heart.

The identification of agents effective against tumors, on the model of the antibiotics, will have to be based on differences in structure and function between the cells of the tumor and normal human cells. Because little is known about the essential differences between tumor cells and normal cells the process of developing antitumor agents is largely empirical, as antibiotic identification was in its early phases. The National Cancer Institute has initiated a large screening program with the goal of identifying agents selectively toxic to tumor cells.

In the industrial production of the beta-lactam antibiotics most if not all of the information needed for synthesis is in the genome of the microorganism; the chemical modifications are relatively

minor. In making other kinds of pharmaceuticals, however, microorganisms mediate only isolated steps, or bioconversions, in a much longer process that relies mainly on nonbiological synthesis. Ordinarily only the information for the isolated steps resides in the genome of the cell, and the DNA specifying these instructions makes up a very small part of the cell's genetic complement.

The most dramatic results of bioconversion methods have been obtained in the manufacture of the steroid hormones. In the early 1930's Edward C. Kendall of the Mayo Foundation and Tadeus Reichstein of the University of Basel isolated cortisone, a steroid secreted by the adrenal gland. About a decade later Philip S. Hench of the Mayo Foundation showed that administration of cortisone could relieve the pain of patients with rheumatoid arthritis. The immediate demand for the hormone was substantial; chemical methods for its synthesis were developed since the potential market was clearly large. The chemical synthesis, however, was elaborate, requiring 37 steps, many of them under extreme conditions. Cortisone made in this way cost $200 per gram.

One of the major complications in the chemical synthesis of cortisone is the need to introduce an oxygen atom at a position in the four-ring steroid structure designated position 11; this step is crucial in determining the physiological activity of the molecule. In 1952 D. H. Peterson and Herbert C. Murray of the Upjohn Company discovered that a strain of the bread mold *Rhizopus arrhizus* is able to hydroxylate progesterone, another steroid, thereby introducing an oxygen atom at position 11. Progesterone is an early intermediate in the synthesis of cortisone; by means of microbial hydroxylation (which is accomplished industrially with microorganisms closely related to *R. arrhizus*) the synthesis was shortened from 37 steps to 11. As a result the price of cortisone was reduced to $6 per gram.

The microbial hydroxylation of progesterone had economic consequences beyond abbreviating the chemical synthesis. The fermentation could be done at 37 degrees Celsius, with water as the solvent and at atmospheric pressure. Reactions under these conditions are much cheaper than those carried out under extremes of temperature and pressure and with solvents other than water, which had been required in the chemical synthesis of cortisone.

Several other uses for microorganisms have since been found in the industrial synthesis of the steroids. The commercially important steroids include the corticosteroids cortisone, hydrocortisone, prednisone and dexamethasone, the androgen testosterone and the estrogen estradiol (the last two used in contraceptives) and spironolactone (a diuretic). The raw materials for all of

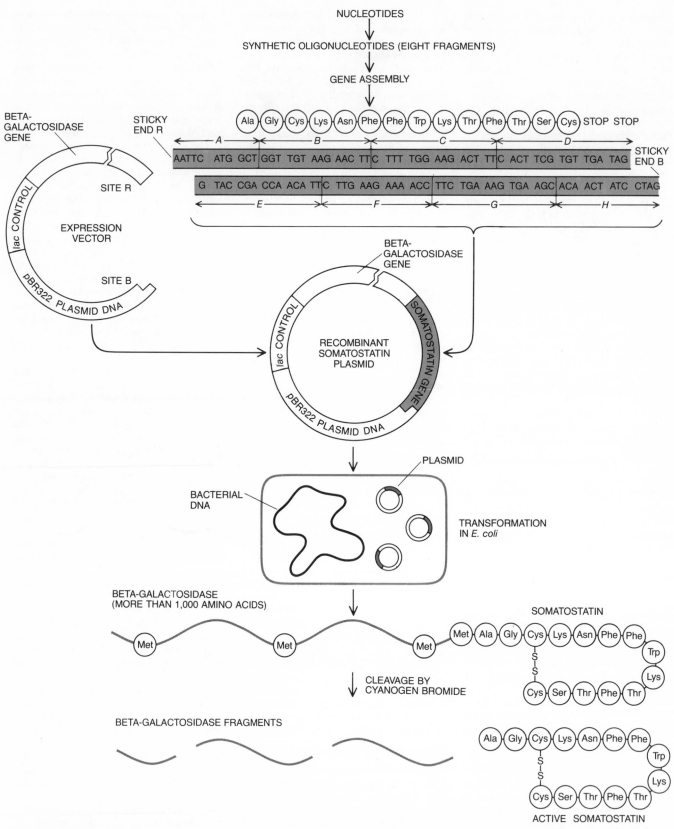

NUCLEOTIDES

SYNTHETIC OLIGONUCLEOTIDES (EIGHT FRAGMENTS)

GENE ASSEMBLY

Ala Gly Cys Lys Asn Phe Phe Trp Lys Thr Phe Thr Ser Cys STOP STOP

BETA-GALACTOSIDASE GENE

STICKY END R

A   B   C   D

AATTC ATG GCT GGT TGT AAG AAC TTC TTT TGG AAG ACT TTC ACT TCG TGT TGA TAG

STICKY END B

G TAC CGA CCA ACA TTC TTG AAG AAA ACC TTC TGA AAG TGA AGC ACA ACT ATC CTAG

E   F   G   H

SITE R

EXPRESSION VECTOR

lac CONTROL

pBR322 PLASMID DNA

SITE B

BETA-GALACTOSIDASE GENE

RECOMBINANT SOMATOSTATIN PLASMID

lac CONTROL

SOMATOSTATIN GENE

pBR322 PLASMID DNA

TRANSFORMATION IN E. coli

PLASMID

BACTERIAL DNA

BETA-GALACTOSIDASE (MORE THAN 1,000 AMINO ACIDS)

Met Met Met

SOMATOSTATIN

Met Ala Gly Cys Lys Asn Phe Phe Trp Lys Cys Ser Thr Phe Thr

CLEAVAGE BY CYANOGEN BROMIDE

BETA-GALACTOSIDASE FRAGMENTS

Ala Gly Cys Lys Asn Phe Phe Trp Lys Cys Ser Thr Phe Thr

ACTIVE SOMATOSTATIN

SYNTHESIS OF SOMATOSTATIN, the first human polypeptide to be produced in bacterial cells, required that the somatostatin gene be inserted into the bacteria by means of an expression vector constructed partly from a plasmid. Somatostatin is a hypothalamic hormone 14 amino acid units long that controls the release of several hormones from the pituitary. The gene was synthesized from eight blocks of single-strand DNA fragments made up of a few nucleotides each, which are indicated in this simplified diagram as *A–H*. The fragments had overlapping complementary sequences to allow for correct assembly of the gene. Of the 52 base pairs in the gene, 42 make up the code for somatostatin; the remainder provide the two "sticky ends" that allow the gene to be inserted into the plasmid and include the in-formation needed for proper expression of the gene and recovery of the hormone. The expression vector was constructed from the plasmid pBR322, to which was added the control region and most of the beta-galactosidase gene from the bacterial *lac* operon. Beta-galactosidase is an enzyme involved in lactose metabolism; the control region contains the regulatory elements needed for expression of the beta-galactosidase gene. The somatostatin gene was inserted into the plasmid next to the beta-galactosidase gene; after the plasmid was introduced into cells of the bacterium *Escherichia coli* the human hormone was synthesized as a short peptide tail at the end of the enzyme. Cleavage with cyanogen bromide freed the hormone, which has been demonstrated to be identical with the molecule in human beings.**

them are the complex alcohols called sterols. There are two common sources of sterols: the production of soybean oil leaves a waste rich in stigmasterol and sitosterol; the roots of the Mexican barbasco plant contain diosgenin.

The first step in steroid production from plant sterols is the degradation of the side chain of the sterol molecule. Several pharmaceutical companies have found it economic to rely on mycobacteria (a group of aerobic, Gram-positive bacteria) to perform this step, which was formerly accomplished nonbiologically. The mycobacteria use sterols as a source of carbon and energy; mutant strains that lack certain enzymes cannot complete the degradation. These mutants are exploited in partial degradations that yield valuable intermediates. Other bacteria modify the steroid nucleus to yield a number of derivatives.

The introduction of these microbiological processes has been of great significance in the manufacture of the steroids, first in making steroid synthesis commercially feasible and later in progressively lowering the unit cost of production. In 1980 the price of cortisone in the U.S. was 46 cents per gram, a 400-fold reduction from the original price. The identification of new uses for steroids (in contraception and in treating hormonal insufficiencies, skin diseases, inflammation and allergies) in conjunction with more efficient production has created a substantial demand for them. World bulk sales of the four major steroids (cortisone, aldosterone, prednisone and prednisolone) amounted to about $300 million in 1978. These are the most important commercial products of bioconversion processes but by no means the only ones.

In the third major class of microbiological processes employed in the manufacture of pharmaceuticals none of the information for the product is initially included in the DNA of the organism; all of it is inserted into the cell. The gene that specifies the structure of the product is first either chemically synthesized or isolated from another organism. By one of several methods the gene is introduced into the cell. Once that is done the existing machinery for gene expression constructs the desired molecule.

The development of methods for transferring a specific gene from one cell to another and for inducing the new host to faithfully and efficiently express the gene has created an enormous potential for the pharmaceuticals industry. A number of companies have been formed to exploit these new capabilities. The immediate outcome has been the production of human polypeptides (short chains of amino acids) by bacteria.

The first human peptide to be synthesized in a bacterial cell was the hypothalamic hormone somatostatin. Somatostatin is one of a group of hormones made in the hypothalamus at the base of the brain; it is then transported in the blood to the pituitary gland, where it acts to inhibit the release of insulin and human growth hormone. Investigators at the City of Hope National Medical Center in Duarte, Calif., and at the University of California at San Francisco chose to work with somatostatin in 1977 mainly because it consisted of only 14 amino acid units. The gene for somatostatin had not been isolated from human cells, but a nucleotide sequence could be deduced from the known order of amino acids in the peptide. A synthetic gene was therefore constructed from blocks of three nucleotides each. Of the 52 base pairs in the synthetic gene, 42 constitute the structural gene for somatostatin. The remaining nucleotides were incorporated to provide suitable "sticky ends" for joining the double-strand DNA fragment to a plasmid, and to facilitate proper expression of the gene and the recovery of the hormone.

The gene was to be inserted into cells of the bacterium *Escherichia coli*. To make the transfer possible the synthetic gene was combined with a plasmid labeled pBR322 and with a segment of the *lac* operon from the *E. coli* genome. (The *lac* operon consists of three physically linked genes involved in lactose metabolism, together with the genetic elements that control their transcription and translation.) The somatostatin gene was inserted close to the end of the gene that codes for the enzyme beta-galactosidase. As a result when the plasmid was placed in the *E. coli* cell, somatostatin was made as a short polypeptide tail attached to the enzyme. By treatment with cyanogen bromide, which breaks proteins into polypeptide fragments at methionine amino acid units, the somatostatin molecules could be recovered. Because the gene had been synthesized chemically it was a simple matter to place a methionine in front of the somatostatin molecule. This approach was necessitated by the fact that when somatostatin is manufactured independently of the enzyme, it is rapidly degraded by bacterial proteins. Somatostatin generated in *E. coli* has been shown to be identical with the natural human hormone. The yield is about 10,000 somatostatin molecules per cell, high enough to encourage further attempts at the industrial production of polypeptides.

Similar techniques have been applied to the bacterial synthesis of human insulin, human growth hormone and interferons. The production of insulin has been even more efficient than that of somatostatin, yielding 100,000 molecules per bacterial cell; insulin is also of greater immediate medical importance than somatostatin. Insulin consists of two polypeptide chains, 21 and 30 amino acid units long. Synthetic genes coding for the polypeptides were constructed in three months' work by Roberto Crea, Adam Kraszewski, Tadaaki Hirose and Keiichi Itakura of the City of Hope National Medical Center. Eighteen fragments of a few nucleotides each were assembled to make the gene for the longer chain, and 11 fragments were joined into a gene for the shorter chain. Each synthetic gene was linked to a plasmid near the end of a beta-galactosidase gene, as in the case of somatostatin. After gene expression and the translation of messenger RNA into protein the two polypeptides were cleaved from the enzyme and linked to form the complete insulin molecule.

The economics of the market for insulin may be fundamentally altered by the application of microbiological techniques. The insulin currently used in diabetes therapy is extracted from the pancreas of cattle and swine. The insulins of these species differ slightly from human insulin in their amino acid sequence; although the animal insulins are effective in controlling the major symptoms of diabetes, they do not prevent some of its ancillary effects, including deterioration of the kidneys and the retina. Moreover, some diabetics are allergic to the animal hormones.

If human insulin manufactured by bacteria proves effective in controlling these pathologies, it will almost certainly gain a substantial share of the world market for insulin, which is now estimated to be about $200 million. Eli Lilly has announced plans to introduce a commercial process for the manufacture of human insulin based on gene transfer. If the yields can be increased to the level of those of other industrial processes that employ *E. coli,* 2,000 liters of fermentation broth could yield 100 grams of purified insulin. This is the amount extracted from some 1,600 pounds of animal pancreatic glands.

The production of somatostatin and insulin by microbiological methods relies on the synthesis of structural genes. With the methods now available, however, it is not economically practical to synthesize genes for peptides longer than about 30 amino acid units, and many clinically important proteins are much larger. Their production calls for the isolation of a natural gene. The starting point in the process is the messenger RNA that encodes the nucleotide sequence for the polypeptide. A complementary DNA copy is made from the messenger RNA by means of the enzyme reverse transcriptase. The double-strand DNA molecule is then replicated many times, and an appropriate vector is chosen to introduce it into bacterial cells.

This approach is beginning to yield results in the production of human growth hormone and the interferons. A deficiency of the pituitary growth hormone results in a form of dwarfism that can be cured by administering the hormone. The hormone is species-specific; its main source has been human cadavers. Growth hormone may have many clinical uses, but the extremely limited availability of the substance has impeded research. Microbiological production may not only increase the commercial availability of the drug but also make it possible to investigate its potential applications. Genentech, one of the companies formed to exploit recombinant-DNA technology, has established a joint venture with Kabi Gen AB, a Swedish company, to make human growth hormone.

Among the other polypeptides under consideration for production by bacteria, the interferons have the most promising applications but also the most uncertain. Interferons seem to have antiviral effects, particularly in preventing (rather than curing) viral infections. They may also have an inhibiting effect on tumor cells. Interferons are synthesized by leukocytes (white blood cells) and by fibroblasts (connective-tissue cells). The interferons available up to now have mainly been extracted from human cells, and the yield is low: two liters of human blood yield one microgram of leukocyte interferon.

Recent developments in recombinant-DNA techniques have made it possible to produce 600 micrograms of leukocyte interferon from a liter of fermentation broth, more than a thousandfold improvement over that obtained from the same volume of blood. The improvement in yield has required a series of modifications of the production process, many of them devised by David Goeddel and his colleagues at Genentech and by Charles Weissman and his colleagues at BioGen. At least four companies are attempting to develop commercial methods of interferon production. One of the most interesting recent developments has been the production of high yields of interferons in yeast cells by fermentation.

Recombinant-DNA methods, particularly those applied to the production of the interferons, may well represent the next great advance in clinical medicine and in the industrial practices of the pharmaceuticals industry. Besides the interferons and insulin the most likely candidates for this form of industrial process include coagulation factors (blood proteins that are required for efficient clotting), enzymes that could serve in replacement therapy for congenital genetic disorders, urokinase (an enzyme that dissolves blood clots), immunostimulants (proteins that trigger immune reactions), antibodies and the antigenic proteins found on the surface of viruses, which might be utilized in the manufacture of vaccines.

The introduction of new genetic methods, however, can hardly represent a departure more radical than the one created by the introduction of antibiotics. The methods associated with antibiotics brought dramatic changes: clinical practice, research and development and industrial production. The manufacture of human polypeptides by microorganisms may lead to the introduction into clinical use of substances, such as the interferons, that have clinical properties as novel as those of the antibiotics. In research and development and industrial production, however, the new methods represent no more than dramatic extensions of genetic methods and fermentation techniques that are already operating on a huge scale in the manufacture of pharmaceuticals.

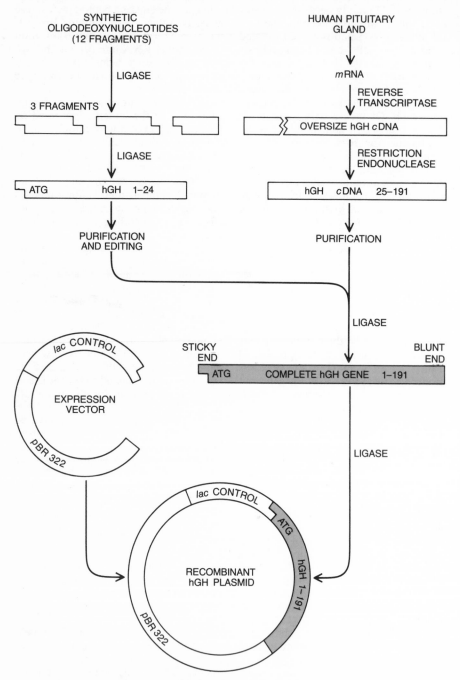

GENE FOR HUMAN GROWTH HORMONE was created by a combination of chemical synthesis and isolation of the natural molecule. Human growth hormone is a polypeptide 191 amino acid units long elaborated by tissues of the pituitary gland. Medical interest in it stems from the fact that its absence leads to a form of dwarfism that can be cured by administration of the hormone. The segment of the gene that codes for the first 24 amino acids of the peptide was constructed chemically from blocks of nucleotides. To obtain the rest of the gene a series of enzymes were used, as this simplified diagram shows. Reverse transcriptase was employed to copy the gene for the hormone from messenger RNA obtained from human pituitary tissues. Restriction endonucleases cut out the needed fragment. DNA ligase was then used to join the natural and the synthetic fragments. The complete gene was inserted into a modified version of plasmid pBR322 incorporating the lac operon. The synthetic part of the growth-hormone gene had been constructed with its own initiation codon (ATG), the group of three bases that provides the signal to start the process of transcription. The hormone could therefore be produced independently in bacterial cells, without the need for attachment to a bacterial protein.

6

The Microbiological Production
of Industrial Chemicals

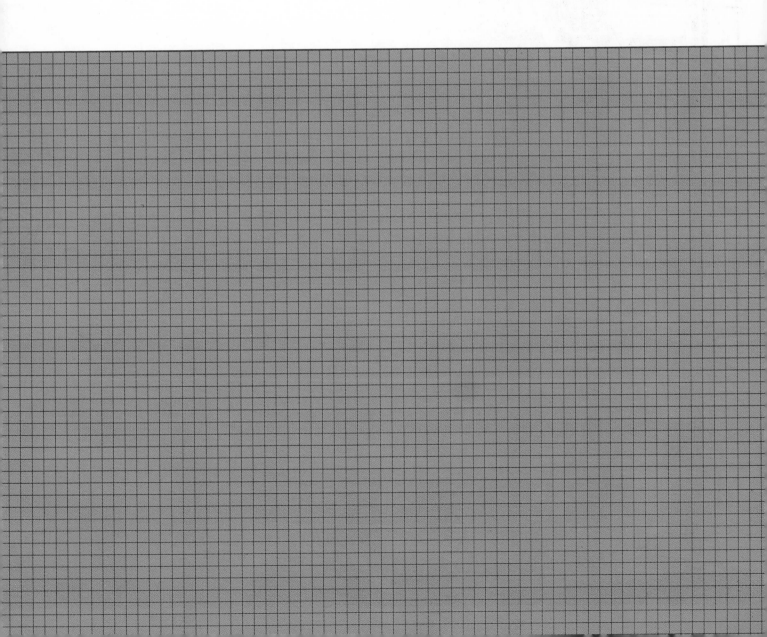

The Microbiological Production of Industrial Chemicals

BY DOUGLAS E. EVELEIGH

Tonnage amounts of many chemicals have traditionally been produced synthetically from fossil fuels. The rising price of petroleum makes fermentative production from other feedstocks increasingly attractive

The synthetic capabilities of microorganisms are not confined to food, drink and pharmaceuticals. Microorganisms also produce industrial chemicals that can either serve as or be employed to make solvents, lubricants, emollients, demulcents, extractants, adhesives, acidulants, plastics, surface coatings, explosives, propellants, gasoline additives, alternative fuels, pesticides, dyes, cosmetics, antifreeze, brake fluid, meat tenderizers, digestive aids, vitamins and flavorings. At various times in the 20th century microbiological fermentation has been the method of choice for the manufacture of citric acid, lactic acid, ethanol, n-butanol and more recently enzymes.

Often an organic substance with industrial applications can be made either biologically or by chemical synthesis. The decision to make it one way or the other is essentially an economic one. A major consideration is the cost of the raw materials. In microbiological fermentation the chief raw material is the growth substrate, which is usually molasses or starch; in chemical synthesis the principal raw material is often petroleum or a derivative of it. The efficiency of the process must be taken into account: What fraction of the substrate is converted into the product and how long does the conversion take? Another factor is the cost of recovering the product from the fermentation medium or from the feedstock in chemical synthesis. One must also weigh the potential value of by-products and the cost of disposing of wastes.

Microorganisms are known to produce some 200 substances of commercial value, but only a few of them are currently made by biological methods in industry; they include ethanol, n-butanol, acetone, acetic acid, citric acid, lactic acid, amino acids and enzymes. Economic considerations suggest that microorganisms will have a larger role in many industries in the 1980's. Because of increases in the price of petroleum the synthetic-chemicals industry no longer enjoys the advantage of an abundant and inexpensive feedstock. Furthermore, the advent of recombinant-DNA techniques makes fermentation more attractive.

Until recently the industrial microbiologist worked with a finite microbial genome; the chief methods available were to search for superior organisms among the products of random mutation or to manipulate growing conditions and thereby modify regulatory pathways to advantage. With the new genetic methods of programming the microbiologist can replace an existing pathway with a new one conducive to higher yields or to faster and more efficient synthesis. To put it another way, he can construct organisms that have novel characteristics and capabilities. Microbiological fermentation in conjunction with the new techniques of genetic programming will contribute significantly to the production of three broad classes of industrial chemicals: enzymes, aliphatic organic compounds and amino acids. I shall take up each of these categories in turn.

Enzymes can catalyze both the making and the breaking of chemical bonds, but they have been exploited commercially mainly to catalyze the decomposition of large molecules such as carbohydrates and proteins. Of the first importance in industrial processes is the specificity of enzymes: each enzyme acts only on a particular substrate molecule.

Because an enzyme is a protein whose functioning depends on the precise sequence of amino acids that make up its structure, large-scale chemical synthesis is impractical. Enzymes either are made by microorganisms grown in culture or are obtained directly from plants and animals. Today costs generally favor the microbiological methods. A major exception is papain, which serves as a digestive aid and a meat tenderizer; it comes from the papaya fruit.

Enzymes have been exploited commercially since the 1890's, when extracts from fungal cells were introduced into brewing to accelerate the breakdown of starch into sugar. Four enzymes are now made on a large scale: protease, glucamylase, alpha-amylase and glucose isomerase. Protease actually encompasses several enzymes that degrade proteins by attacking peptide bonds. The main industrial protease, obtained from the bacterium *Bacillus licheniformis,* serves chiefly as a cleaning aid in detergents. Proteases from other bacteria and fungi are employed on a smaller scale as digestive aids in animal feed and as meat tenderizers. The amylases are a family of enzymes that break down starch first into short chains of glucose molecules and then into free glucose. Glucose isomerase converts glucose into its stereoisomer fructose, a sweetener.

Industrial fermentation is responsible for making some 1,270 tons per year of

OUTDOOR FERMENTATION TANKS contain microorganisms that convert sugar into the amino acids glutamic acid and lysine. Glutamic acid serves as a flavor enhancer in the form of the salt monosodium glutamate (MSG). Lysine is an amino acid that is essential to the nutrition of man and nonruminant animals but that cannot be synthesized by them; it is added to animal feed. Each tank holds 63,420 gallons and is roughly 100 feet high. The photograph shows seven of some 20 identical tanks at the plant of Kyowa Hakko Kogyo Co., Ltd., at Hofu in Japan. The tanks, which were built in the early 1970's, are the largest ones at the plant that make amino acids. Kyowa Hakko annually produces at least 20,000 tons of MSG and 10,000 tons of lysine. The amino acids are made by an aerobic process; from each tank the pipe that appears to pass through a funnel expels exhaust. Amino acids constitute a class of industrial chemicals that can be made microbially; two other classes are enzymes and aliphatic organic compounds.

the four enzymes: 530 tons of protease, 350 tons of glucamylase, 320 tons of alpha-amylase and 70 tons of glucose isomerase. Sales of all enzymes worldwide now amount to about $300 million. The industry is dominated by European companies, with Novo Industri in Denmark and Gist-Brocades NV in the Netherlands having 60 percent of the world market. The industry is expected to grow in the next decade as recombinant-DNA techniques are applied to the microbiological production of enzymes. Because an enzyme is a direct product of a gene the yield can be improved by introducing multiple copies of the gene into the DNA of the organism, by maximizing the expression of the gene through insertion of regulatory sites in the DNA called promoters and by facilitating the secretion of the enzyme from the cell.

The process that has most clearly demonstrated the value of microbiologically synthesized enzymes is the conversion of starch into the sweetener high-fructose corn syrup, which is rapidly replacing sucrose in soft drinks. Although the conversion was introduced only recently, it is already yielding more than two million tons of high-fructose corn syrup per year. The conversion has three steps, in which the feedstock is acted on successively by alpha-amylase, glucamylase and glucose isomerase.

The cost of making fructose sweeteners depends strongly on the efficiency with which the enzymes can be obtained. Bunji Maruo and his co-workers at Nihon University have increased the yield of alpha-amylase from *Bacillus subtilis* almost 200 times by combining the classical method of mutation and selection with the technique of genetic recombination. They have found a number of regulatory steps controlling the synthesis of alpha-amylase that act synergistically to give enhanced yields in the selected strains of *B. subtilis*.

Recombinant-DNA techniques have also been applied to the production of a heat-stable alpha-amylase. *B. subtilis* grows at room temperature and the alpha-amylase it synthesizes is readily denatured by heat. If the enzyme could act at a higher temperature, the catalytic conversion of starch into glucose would proceed at a higher rate. One way of making such an enzyme would be to insert the alpha-amylase gene from a thermophilic bacterium into *B. subtilis*. Thermophilic bacteria live under conditions of high temperature and make enzymes that are resistant to inactivation by heat. They cannot be advantageously exploited to produce alpha-amylase, however, because their genetic structure is poorly understood. Shoji Shinomiya and his co-workers at the University of Tokyo have demonstrated that increased yields of thermostable alpha-amylase can be obtained by the insertion of an amylase gene from a thermophilic bacterium into *B. subtilis*.

Another way of increasing the efficiency of fructose manufacture would be to condense the present three steps into one step. This might be accomplished by incorporating genes for alpha-amylase, glucamylase and glucose isomerase into a single microorganism. Starch would then be converted into high-fructose corn syrup in a single fermentation vessel.

A realm in which enzymes derived from microorganisms may soon make an important contribution is the $50-billion plastics industry. Several plastics are made by the polymeriza-

tion of alkene oxides, that is, oxides of straight-chain carbon compounds in which at least one of the bonds between carbon atoms is a double bond. The alkene oxides are now made by chemical synthesis from petrochemical feedstocks. An elegant enzymatic approach to the synthesis of alkene oxides was recently proposed by Saul L. Neidleman of the Cetus Corporation. Peter J. Farley, the president of Cetus, expects that the enzymatic synthesis of propylene oxide will be introduced commercially before the end of the decade and will reach annual worldwide sales of between $2 billion and $3 billion.

The enzymatic synthesis of alkene oxides from alkenes relies on three enzymes: pyranose-2-oxidase from the basidiomycete fungus *Oudemansiella mucida,* a haloperoxidase from the fungus *Caldariomyces* or other sources and an epoxidase from *Flavobacterium*. In the first step of the synthesis pyranose-2-oxidase promotes the formation of hydrogen peroxide (H_2O_2), with glucose serving as the substrate and energy source. In the second step, which is mediated by the haloperoxidase, the hydrogen peroxide combines with an alkene and a halogen ion (fluoride, chloride or bromide) to form an alkene halohydrin: an alkene bonded to a hydroxyl group (–OH) and a halogen. In the final step, mediated by the epoxidase, the hydrogen of the hydroxyl group and the halogen ion are stripped away, leaving the alkene oxide.

The enzymatic production of alkene oxides could have economic advantages over chemical synthesis. A halogen ion can be supplied by a simple salt such as sodium chloride and is therefore less expensive than an elemental halogen, which is required in the chemical synthesis. The enzymatic system can also generate by-products such as fructose and gluconic acid. (The latter is added to dishwater detergents because it prevents the precipitation of calcium and magnesium salts, which can leave spots on glass surfaces.) The production of fructose from glucose in Neidleman's proposal is of major importance in view of the growing use of fructose as a sweetener. In Neidleman's scheme the conversion of glucose first into glucosone and then into fructose has a 100 percent yield, which compares favorably with the maximum yield of about 50 percent in the enzymatic conversion of starch into high-fructose corn syrup.

Another advantage of the enzymatic synthesis of alkene oxides is its flexibility: by changing the substrate on which the haloperoxidase acts the process can be adjusted to yield different alkene oxides, such as propylene oxide for the plastic polypropylene and ethylene oxide for the plastic polyethylene. Still another advantage is the absence of pollutants because the halogen can be recy-

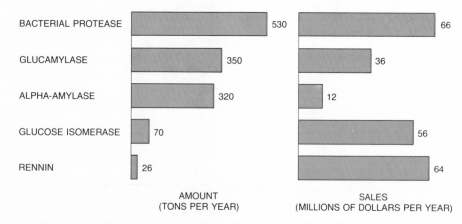

ENZYME	AMOUNT (TONS PER YEAR)	SALES (MILLIONS OF DOLLARS PER YEAR)
BACTERIAL PROTEASE	530	66
GLUCAMYLASE	350	36
ALPHA-AMYLASE	320	12
GLUCOSE ISOMERASE	70	56
RENNIN	26	64

WORLDWIDE SALES OF ENZYMES were $300 million in 1980. Here the annual tonnage and the annual sales are given for five enzymes that are made on a large scale by microbiological methods. Bacterial protease, which degrades proteins by cleaving peptide bonds, has commercial value chiefly as a cleaning aid. The enzymes alpha-amylase, glucamylase and glucose isomerase serve mainly to convert starch into the sweetener high-fructose corn syrup, which is replacing sucrose in soft drinks. Amylases break down starch to yield glucose; glucose isomerase converts glucose into fructose. Rennin is employed in making cheese. It can be extracted from the fourth stomach of a calf or a cow or it can be made microbiologically. Data came from the Office of Technology Assessment and from J. Leslie Glick of the Genex Corporation.

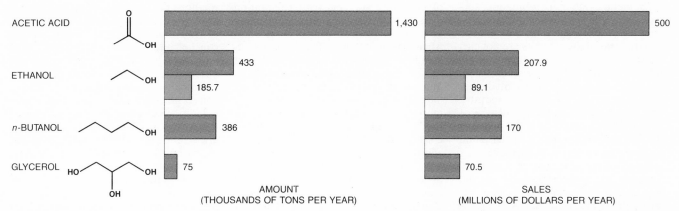

ALIPHATIC ORGANIC COMPOUNDS, apart from methane, had total sales in the U.S. of $3 billion in 1980. The aliphatic compounds include solvents and organic acids. Shown here are four that are made in large quantities: acetic acid, ethanol, n-butanol and glycerol. Excluded from this accounting are the ethanol made for alcoholic beverages and the acetic acid employed as vinegar. All four compounds can be made by biological means but only ethanol is now made that way in industry, and 70 percent of the industrial ethanol is synthesized nonbiologically from petroleum derivatives. The colored bars mark biological syntheses, the gray bars nonbiological ones. The aliphatic industry is expected to adopt fermentation more generally because of the cost of petroleum, the possibility of exploiting thermophilic (heat-loving) bacteria and the prospect of new feedstocks. Data are from the U.S. International Trade Commission.

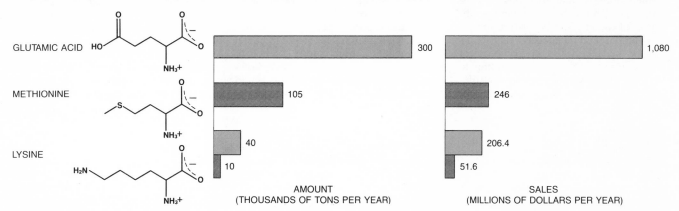

WORLDWIDE SALES OF AMINO ACIDS were $1.7 billion in 1980. The three amino acids shown are the ones now made in the largest quantities: glutamic acid, methionine and lysine. Glutamic acid is made by fermentation. Like lysine, methionine is a nutritionally essential amino acid that is made commercially as an animal-feed additive. Methionine is manufactured by chemical synthesis, but 80 percent of the lysine is produced biologically. Each amino acid has two isomers, only one of which participates in biological reactions. Fermentation yields only the biologically active isomer; in chemical synthesis half of the yield is the inactive one. This specificity makes biological means more efficient, but it has not always been possible to exploit it. With increased understanding of cellular metabolism all industrially valuable amino acids may soon be made by fermentation. Data are from the Office of Technology Assessment and from Glick.

cled. Moreover, the enzymatic system will undoubtedly be improved by recombinant-DNA techniques. To begin with, genetic programming should be able to increase the yield of the three enzymes from their microbial sources. An intriguing but remoter possibility is to enhance the performance of the enzymes by modifying their active sites.

Commercially produced enzymes are playing an increasing part in medical diagnosis. For example, the enzyme cholesterol oxidase is employed to monitor the level of cholesterol in blood serum, and the enzyme uricase serves to monitor the level of uric acid.

Recombinant-DNA technology itself requires certain enzymes such as restriction endonucleases for cutting open DNA, and ligase for resealing the cut ends. The companies that work with recombinant DNA clearly have an interest in the inexpensive production of these enzymes. Ronald W. Davis and his co-workers at the Stanford University School of Medicine have achieved a 500-fold increase in the yield of ligase by inserting multiple copies of the ligase gene into Escherichia coli. The enzymes required by recombinant-DNA techniques have been beneficial in medical diagnosis. For example, the prenatal detection of sickle-cell anemia can be done by applying a restriction endonuclease to the DNA of fetal cells in the amniotic fluid. This method has none of the hazards of the traditional diagnostic technique, in which blood is drawn from the fetus.

Recombinant-DNA techniques have much to offer the enzyme industry apart from the simple enhancement of enzyme yield. Direct modification of the gene could yield enzymes with greater specific activity and greater thermal stability. Eventually the rational design and synthesis of enzymes will be possible. Other factors of economic importance in the production of enzymes include the more efficient use of the feedstock and other raw materials, inhibition of the synthesis of unwanted enzymes and the more efficient recovery of enzymes from dilute solutions. Each of these factors can be addressed by genetic-programming techniques.

The operational efficiency of the fermenter might also be improved by genetic-programming techniques. When the fermenter is a fungus, the long filaments of the fungus often become tightly intertwined, like a clump of spaghetti, with the result that they cannot effectively take in nutrients from the growth medium. If the fungus could be genetically modified so that it grew not as filaments but as single cells, it could consume more nutrients and ferment more efficiently.

The second major class of industrial chemicals is made up of aliphatic organic compounds, which are distinguished by the absence of benzene rings and similar structures. The aliphatic substances with industrial applications can be broadly divided into two categories: solvents and organic acids. The solvents include ethanol, *n*-butanol, acetone and glycerol; the organic acids include acetic acid, citric acid and lactic acid. In general the solvents are not currently made by biological means on an industrial scale, although *n*-butanol, acetone and glycerol were once made that way. Nevertheless, the solvent industry may return to fermentation because of the cost of petrochemicals, the prospect of exploiting thermophilic bacteria and the availability of new feedstocks.

Thermophilic bacteria grow rapidly in the range of temperatures between 60 and 75 degrees Celsius. Their chief advantage over microorganisms that grow at a lower temperature is their faster metabolism. Another benefit is that the fermenter need not be cooled much in order to remove the heat given off by the metabolism of the bacteria. In addition, when the solvent stream issues from the fermenter at a high temperature, less energy is needed for the subsequent purification of the product by distillation.

Abundant and inexpensive feedstocks will be needed if the microbiological production of solvents is to compete with chemical synthesis from petroleum derivatives. In the earlier methods

of preparing solvents by fermentation the substrate was sugar from sugarcane or beet molasses or starch from corn, wheat, rye or cassava. The price of sugarcane, molasses and starch is subject to wide fluctuations, and so it would be difficult to base a large fermentation industry on these substrates. Furthermore, these materials are needed for food. The alternative substrates being considered include cellulose, methanol and organic wastes.

Cellulose and related polymers are a major constituent of almost all plant materials and thus represent a renewable feedstock. A promising source of cellulose for the production of solvents is wood. It is widely available and has a stable, low price compared with that of sugar and starch. Wood has three structural components: lignin and the polysaccharides cellulose and hemicellulose. For the fermentation of wood to compete with the synthetic-chemicals industry all three components must be utilized. Lignin presents no problem: it can be burned as a high-caloric fuel. Efficient methods must be developed, however, for fermenting the cellulose and the hemicellulose. I shall discuss these

methods below when I describe the production of ethanol.

Methanol, another possible substrate, can be made from coal that has been converted into synthesis gas. Methanol has a single carbon atom, and subtle biochemical pathways are required for its conversion into more complex molecules in which carbon atoms are bonded together. Only a few microorganisms are able to get all their carbon from single-carbon compounds such as methanol, and so any commercial process based on a methanol substrate will have to rely on one of those organisms. Appropriate regulatory mechanisms and biochemical pathways will have to be genetically programmed into the organisms. Imperial Chemical Industries in Britain has grown the bacterium *Methylophilus methylotrophus* on methanol, having first deleted an energy-dependent pathway by methods of genetic engineering. The bacterium produces a single-cell protein that is sold as an animal feed called Pruteen.

The use of organic wastes as a substrate for the production of solvents would have the secondary benefit of disposing of noxious materials. The wastes

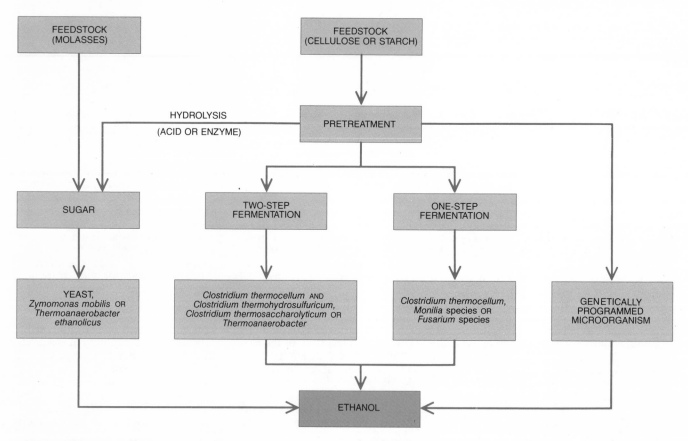

BIOLOGICAL SYNTHESIS OF ETHANOL is now done by essentially the same method employed in making alcoholic beverages, but other methods are being considered. The substrate is now either crude sugar (from sugarcane or beet molasses) or starch (from corn, wheat, rye or cassava) that has been converted into sugar. A yeast has usually been the fermenting organism, but the bacteria *Zymomonas mobilis* and *Thermoanaerobacter ethanolicus* may be more efficient. The prices of crude sugar and of starch fluctuate widely, so that it would be difficult to base a large fermentation industry on these substrates. Furthermore, these materials are needed as food. Three oth-

er strategies rely on cellulose and related polymers of wood, which are abundant, renewable and inexpensive. The cellulose can be fermented in either two steps or one step. In the two-step process one microorganism breaks down the cellulose into its component sugar units, which are subsequently fermented into ethanol by another microorganism. In the one-step process a single microorganism both breaks down the cellulose and ferments the resulting sugar solely into ethanol. Microorganisms that have this capability are apparently rare, and so the fourth strategy is to genetically program a yeast or a bacterium so that it converts cellulose into ethanol in one operation.

are a complex mixture of substances, some of which are toxic. In order to ferment the wastes a microorganism must be found that is resistant to any toxins present and that can consume several component substances of the mixture. A promising family of bacteria consists of the pseudomonads. Ananda M. Chakrabarty of the General Electric Company genetically programmed *Pseudomonas putida* so that it utilizes naphthalene, xylene, alkanes and camphor. The suggestion that the bacterium could degrade oceanic oil spills exaggerates its capabilities, but it will prove useful in several industrial settings, such as waste-disposal ponds, in which the temperature and other environmental factors can be controlled.

Ethanol is one of the most important organic chemicals in industry. In the U.S. 619,000 tons were produced last year, for total sales of $297 million. (These amounts do not include the ethanol made for alcoholic beverages.) Ethanol is employed as a solvent, an extractant and an antifreeze. Moreover, it is a substrate for the synthesis of other organic compounds that serve as solvents, extractants, dyes, pharmaceuticals, lubricants, adhesives, detergents, pesticides, plasticizers, surface coatings, cosmetics, explosives and resins for the manufacture of synthetic fibers.

Ethanol can be made either synthetically from petrochemical feedstocks or biologically by the yeast *Saccharomyces cerevisiae* or other microorganisms. In a common chemical process ethylene derived from petroleum or natural gas is converted at a high temperature into ethanol by the addition of water and in the presence of certain catalysts. In the microbiological process a yeast secretes ethanol as a by-product of fermenting either crude sugar or starch that has been converted into sugar. At the beginning of the 20th century ethanol was produced on a large scale by fermentation. In recent years 70 percent of the ethanol made in the U.S. has been made by chemical synthesis, chiefly because of the cost of sugar and starch. The rising cost of petroleum, however, is pushing the industry back to fermentation.

Indeed, the balance of costs has shifted to such an extent that the fermentative manufacture of ethanol as an alternative to gasoline is now well established. Brazil plans to replace gasoline with ethanol by the 1990's. In the U.S. gasohol, a 9:1 blend of gasoline and ethanol, has been enthusiastically received in the Middle West. Replacing gasoline with gasohol throughout the country would require at least 12 billion gallons of ethanol per year; current production is about .3 billion gallons, of which only a third is employed as a fuel. Nevertheless, the price of ethanol made by fermentation is comparable to the price of gasoline.

If ethanol is to serve as a fuel, the requirement of an abundant and inexpensive substrate is particularly pressing. In a corn glut, such as the current one in the U.S., excess grain could be used as the substrate, but it would yield at most perhaps two billion gallons of ethanol per year.

A more abundant substrate is wood, but the biological synthesis of ethanol from wood is appreciably more complicated than that from grain. The wood must be pretreated in various ways and its cellulose and hemicellulose must be separated. Then the cellulose can be fermented either in two steps or in one step. In the two-step process the cellulose, which is a polymer of glucose, is broken down into its component sugar units, which are fermented into ethanol. The process is efficient, but its compound nature adds to the cost. The alternative approach is to break down the cellulose and ferment the resulting sugar in one operation. Microorganisms with the full complement of enzymes needed to bring about this transformation are apparently rare. As a result investigators are trying to genetically program an ethanol-producing yeast or the bacterium *Zymomonas* so that it can degrade cellulose.

Similar efforts center on the hemicellulose xylan, which is a polysaccharide having a structure somewhat different from that of cellulose; it is composed of the pentose sugar xylose. Xylan can make up as much as 30 percent of the mass of wood. Groups of investigators led by Henry Schneider of the National Research Council of Canada and George T. Tsao of Purdue University have independently shown that if xylose is converted into xylulose, it can be fermented by yeast to yield ethanol. The conversion of xylose into xylulose is mediated by the enzyme xylose isomerase, and so investigators are trying to incorporate into yeast the gene for xylose isomerase.

What microorganism produces the most ethanol? Yeasts, of course, have been the traditional choice, but workers at the University of New South Wales and at Rutgers University have discovered that the bacterium *Zymomonas mobilis,* found in palm wines and in the Mexican beverage pulque, ferments sugar twice as fast as yeast does. Ultimately thermophilic bacteria will probably prove to be the most efficient fermenters. Even their efficiency is subject to improvement by recombinant-DNA techniques, which can serve to increase the amount of certain enzymes in the cell or to replace one enzyme with another that has a higher specific activity.

A major limitation on the fermentative production of ethanol and other solvents is the capacity of the microorganism to tolerate the solvent. Not much is known about what makes a microorganism resistant to ethanol, but tolerance seems to be linked to the fraction of fatty acids in the cell membrane that are chemically unsaturated, or deficient in hydrogen. One possible explanation

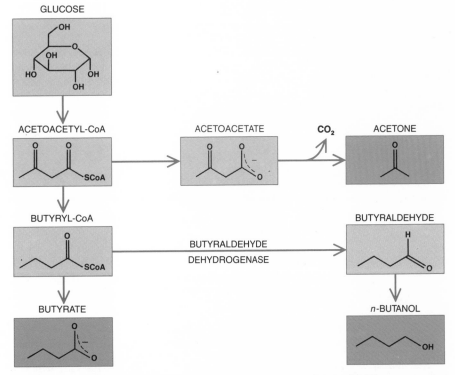

SYNTHESIS OF *N*-BUTANOL can be accomplished by microbiological means, but such methods have not been adopted because the biochemical pathways are not well understood. The pathways shown here represent the fermentation of glucose into *n*-butanol by the bacterium *Clostridium acetobutylicum.* Butyrate (the salt of butyric acid) is formed initially, followed by a metabolic rerouting to the production of acetone and *n*-butanol. The mechanism that controls this is not known. Another critical factor is the toxicity of *n*-butanol to the bacterium.

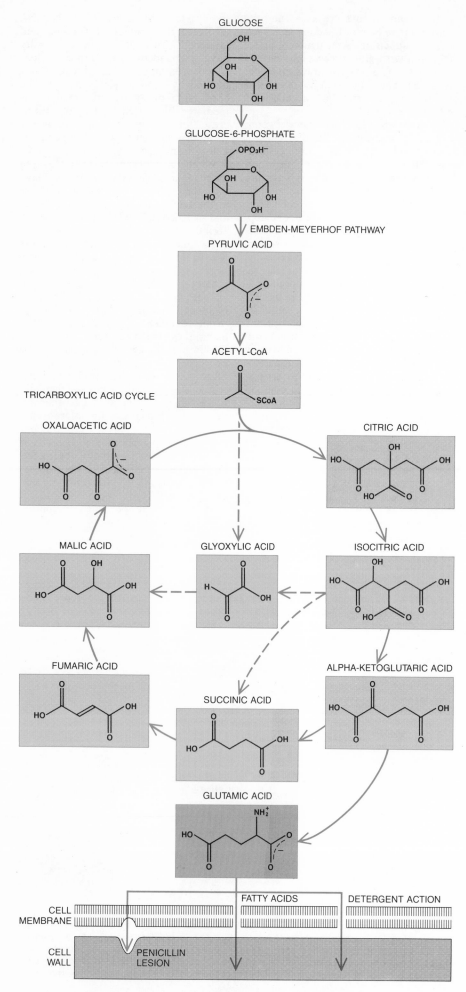

GLUCOSE

GLUCOSE-6-PHOSPHATE

EMBDEN-MEYERHOF PATHWAY

PYRUVIC ACID

ACETYL-CoA

TRICARBOXYLIC ACID CYCLE

OXALOACETIC ACID

CITRIC ACID

MALIC ACID

GLYOXYLIC ACID

ISOCITRIC ACID

FUMARIC ACID

ALPHA-KETOGLUTARIC ACID

SUCCINIC ACID

GLUTAMIC ACID

FATTY ACIDS

DETERGENT ACTION

CELL MEMBRANE

CELL WALL

PENICILLIN LESION

is that unsaturated fatty acids make the membrane more permeable to ethanol and thereby reduce the intracellular concentration.

Another organic solvent is *n*-butanol. (The *n* stands for "normal" and signifies that the molecule is a straight chain of carbon atoms rather than a branched chain.) It is employed extensively in the manufacture of plasticizers, brake fluids, gasoline additives, urea-formaldehyde resins, extractants and protective coatings. Today virtually all *n*-butanol is made by chemical synthesis, but a biological pathway has long been known. In 1912 Chaim Weizmann, who was then working at the University of Manchester, developed a bacterial fermentation culture for the production of *n*-butanol, from which he synthesized butadiene for synthetic rubber. Acetone is a by-product of the fermentation, and in World War I there was a great demand for acetone as a solvent in the manufacture of the explosive cordite. After the war the demand for acetone decreased but the need for *n*-butanol increased.

The fermentation devised by Weizmann was based on the conversion of starch by the bacterium *Clostridium acetobutylicum* or the conversion of sugar by *Clostridium saccharoacetobutylicum*. The production of *n*-butanol by fermentation declined in the 1940's and the 1950's mainly because the price of

PRODUCTION OF AN AMINO ACID by biological methods depends on strategies for altering the metabolism of the cell and for promoting the excretion of the product from the cell. Shown here is the conversion of glucose into glutamic acid (or the salt MSG) by *Corynebacterium*. Glucose is converted into pyruvate by the Embden-Meyerhof pathway (glycolysis), a fundamental mechanism by which a cell derives energy from glucose. The pyruvate then enters the Krebs cycle, or tricarboxylic acid cycle, in which further oxidation releases more energy. The cycle is short-circuited by a low level of the enzyme alpha-ketoglutarate dehydrogenase, which promotes the conversion of alpha-ketoglutaric acid into succinic acid, and by a high level of the enzyme glutamate dehydrogenase, which encourages the conversion of alpha-ketoglutaric acid into glutamic acid. At the same time the glyoxylate cycle, represented by the broken arrows, is activated to produce more energy. While the cell is growing the glyoxylate cycle competes with the glutamic acid shunt for alpha-ketoglutaric acid. Eventually all the substrate is shunted into glutamic acid. There are several strategies for inducing the cell to excrete glutamate in large quantities. If the cell is deprived of the vitamin biotin, the membrane develops leaks, allowing more glutamate to pass through. The membrane can also be effectively modified by adding to the growth medium a saturated fatty acid or a detergent. Addition of penicillin to growth medium causes lesions in the cell wall, through which large quantities of glutamic acid can be excreted.

petrochemicals dropped below that of starch and sugar substrates such as corn and molasses. Today *n*-butanol is made by fermentation only in South Africa, where petroleum is scarce because of the international embargo. The substrate is molasses, and coal is also required as a source of energy; the spent bacteria are recovered and sold as a feed for ruminants. In response to the rising cost of petrochemicals, industry in other countries is reexamining fermentation as a source of *n*-butanol.

One impediment to the large-scale adoption of microbiological methods for *n*-butanol synthesis is a lack of understanding of metabolic pathways. Another critical factor is the toxicity of *n*-butanol to the bacterium. The first industrial strains of *C. acetobutylicum* fermented at most 3.8 percent of the starch substrate to yield 1.2 percent *n*-butanol. The bacteria are killed by higher concentrations of this product of their own metabolism. Investigators have not had much success in selecting mutant strains with greater tolerance to *n*-butanol. In laboratory evaluations a tolerance of 2.85 percent has been achieved by adding activated charcoal to the fermentation medium, but it is not known whether this can be done on a large scale. Once it is understood what makes a substance toxic to an organism, recombinant-DNA techniques may suggest a strategy for increasing the organism's resistance.

Glycerol serves as a lubricant in the manufacture of pet food, baked goods, candy, icing, toothpaste, adhesives, glue, cork board, cellophane and certain kinds of paper. As an emollient and demulcent it is an ingredient in many pharmaceuticals and cosmetics. As a solvent it has been added to extracts, other flavorings and food col-

FEEDBACK INHIBITION is an intracellular regulatory mechanism that must be overcome in the commercial production of certain amino acids such as lysine. The upper diagram shows the metabolic pathway for the fermentative production of lysine by *Corynebacterium glutamicum*. The colored lines represent feedback inhibition. Lysine and threonine are both made by the bacterium, and their simultaneous accumulation inhibits the action of the first enzyme in the pathway, aspartate kinase, and hence inhibits the further production of lysine. In a mutant strain that lacks the enzyme homoserine dehydrogenase the metabolic steps leading to the synthesis of threonine are eliminated; the missing steps are in the gray region. The mutant needs threonine in order to grow, but it is added slowly so that it and the accumulated lysine do not trigger feedback inhibition. In the absence of the inhibitory mechanism synthesis of lysine proceeds at a maximum rate. The lower diagram shows the metabolic pathways for the production of lysine by *Escherichia coli*. The mechanism of feedback inhibition is more complex, as the colored lines indicate, and so *E. coli* is not employed commercially to make lysine.

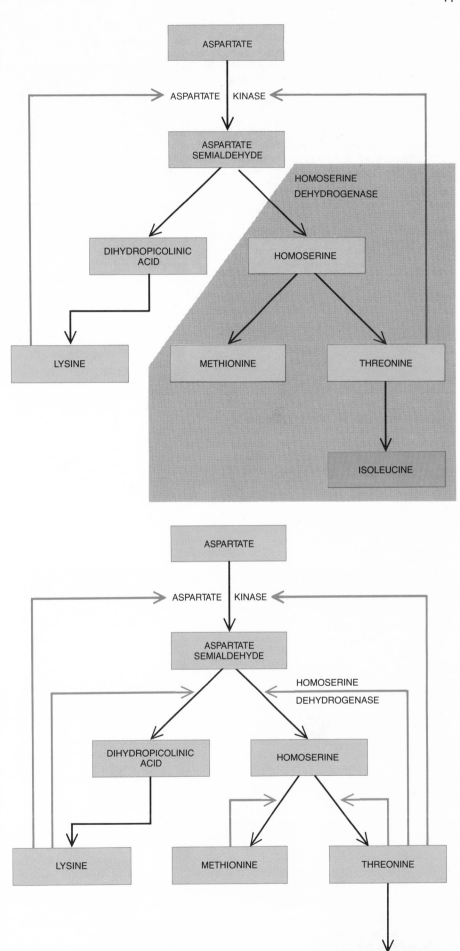

ors. Esters of glycerol figure in the manufacture of explosives and propellants.

During World War I, Germany made large amounts of glycerol by fermentation for use in explosives. The process consisted in adding sodium sulfite to an ethanol fermentation culture. The sulfite interferes with the synthesis of ethanol by combining with an intermediate molecule. In the resulting diversion of the metabolic pathways glycerol becomes the major end product. Since World War I glycerol has been made commercially not by fermentation but by the saponification of fats and by synthesis from propylene and propane. The microbiological production of glycerol is being considered again mainly because of the discovery of yeasts that can synthesize it without the need for sodium sulfite or other steering agents.

Of the industrial organic acids acetic acid is the most important. In the U.S. more than 1.4 million tons are produced each year, with a value of $500 million. (Excluded from these amounts is the acetic acid that serves as vinegar.) The acid also has applications in the manufacture of rubber, plastics, acetate fibers, pharmaceuticals, dyes, insecticides and photographic materials. In Japan acetic acid is a substrate for the fermentative production of amino acids.

Acetic acid can be formed by the microbiological oxidation of ethanol, but except in making vinegar the process is not currently competitive with the chemical synthesis, which is based on the carbonylation of methanol. In the U.S. promising work is being done on the fermentation of cellulose into acetic acid by a thermophilic bacterium. Another possible approach is the conversion of hydrogen and carbon dioxide into acetic acid by the bacteria *Acetobacterium woodii* and *Clostridium aceticum*. The development of such a technology will depend on deciphering the genetics of these little-studied bacteria.

Citric acid, an essential ingredient in foods, is produced efficiently from molasses by *Aspergillus niger*. The world market for citric acid is 175,000 tons per year, for sales of $259 million. The fermentation would be more economical if it were based on a less expensive substrate such as cellulose. This could be achieved by transferring to *A. niger* genes for enzymes that easily decompose cellulose. The commercial fermentative production of certain organic acids such as acrylic acid, which is used in plastics, and N-acetyl *p*-aminophenol, which is marketed as the aspirin-substitute Tylenol, will benefit less from genetic programming than it will from a better understanding of the regulatory mechanisms favoring their synthesis.

Lactic acid, which serves as an acidulant in food and a mordant in textiles and is employed in electroplating, electropolishing and the manufacture of plastics, was the first organic acid to be made commercially by fermentation. In the U.S. and Europe about 40,000 tons are made each year, with sales of $56 million. Virtually all the lactic acid produced in the U.S. is now made by chemical synthesis, whereas only half of it is made that way in Europe. Lactic acid is fermented efficiently from glucose by the bacterium *Lactobacillus delbrueckii*, but the recovery of the acid from the culture is expensive.

I turn now to amino acids, the small molecules that are assembled to make up proteins. Of the 20 amino acids that are generally incorporated into proteins eight cannot be synthesized in man; among these eight essential amino acids lysine and methionine are particularly important to nutrition because most cereal grains are deficient in them. Lysine and methionine are therefore made commercially as animal-feed additives. Methionine is produced synthetically, whereas only 20 percent of all lysine is made that way; the remaining 80 percent is made by fermentation. Another industrially important amino acid is glutamic acid, which is employed as a flavor enhancer in the form of a salt, monosodium glutamate (MSG); it is made only by microbiological means. The fermentative production of 40,000 tons of lysine per year and 300,000 tons of MSG per year is a success story in the microbiological production of industrial chemicals.

In general an amino acid can be made more efficiently by fermentation than by chemical synthesis whenever the intracellular metabolic pathways governing its production are understood and are not too complex. For each amino acid there are two isomers, or mirror-image arrangements of the molecule. With few exceptions only one of the isomers participates in biological reactions. The microbiological production of an amino

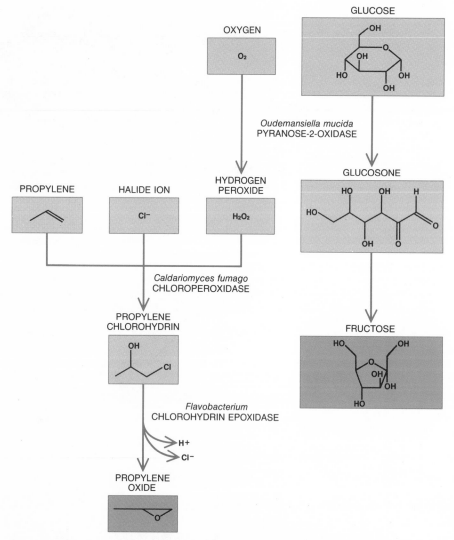

ENZYMATIC SYNTHESIS OF ALKENE OXIDES, which are raw materials in the plastics industry, has been proposed by Saul L. Neidleman of the Cetus Corporation. Plastics such as polypropylene and polyethylene are made by the polymerization of alkene oxides, which are now synthesized from petrochemical feedstocks. The enzymatic synthesis relies on three enzymes: pyranose-2-oxidase, a haloperoxidase and an epoxidase. The fungal and bacterial sources of the enzymes are indicated in the diagram. The system can generate valuable by-products.

acid yields solely the biologically active isomer, whereas the chemical synthesis yields equal amounts of both spatial arrangements. In other words, half of the yield in the chemical synthesis is biologically inactive. Moreover, methods for separating the isomers are expensive and in some instances are not known at all. I suspect that with increased understanding of cellular metabolism all commercially valuable amino acids will be made solely by fermentation.

MSG is produced on a large scale by cultures of *Corynebacterium glutamicum* and *Brevibacterium flavum*. In the U.S. the flavoring industry consumes 30,000 tons of MSG annually; a third of this amount is imported from Japan and from South Korea. The principal substrate has been glucose; alternatives are the *n*-paraffin fractions of petroleum, which were employed in the late 1960's when they were plentiful and cheap, and acetic acid, which is inexpensive and gives rise to fewer wastes than glucose.

An important factor in the commercial production of MSG (and other amino acids) is inducing the cell to excrete it in large quantities. One strategy adopted for this purpose is to grow *Corynebacterium glutamicum* in a medium with less than the optimum amount of the vitamin biotin. The cell membrane then becomes deficient in phospholipids; as a result it develops leaks and allows more MSG to be excreted. If the growth medium has a high level of biotin, the membrane must be modified in another way, such as by the addition of a saturated fatty acid or a detergent. Also effective is the addition of penicillin, whose antibacterial mode of action is to inhibit the synthesis of peptidoglycan in the cell wall.

When the mechanisms governing the excretion of MSG are understood, it should be possible to apply recombinant-DNA techniques to *C. glutamicum* in order to create leaky membranes.

Such genetically altered strains will yield high levels of MSG without a requirement for precise regulation of the growth conditions and without the need for expensive additives such as saturated fatty acids, detergents or penicillin.

The yield of an amino acid can also be increased by the development of a bacterial strain in which a regulatory mechanism that ordinarily limits the production of an amino acid is circumvented. In *C. glutamicum*, for example, lysine and threonine are both products of the same sequence of synthetic events, and the simultaneous presence of both amino acids inhibits an enzymatic step early in the pathway and hence impedes the further production of lysine. In a certain mutant strain the steps leading to the synthesis of threonine are blocked. The mutant bacteria need threonine to grow, but it is added slowly so that it and the accumulated lysine do not inhibit the further production of lysine. With the loss of regulation the synthesis of lysine proceeds at the maximum rate. In another approach no steps are blocked but the mechanism of feedback inhibition itself is disabled, and so the lysine and threonine are allowed to accumulate.

The outlook for the amino acid industry is bright because of new and expanded markets. The world demand for animal protein is high, and so is the need for lysine and methionine as feed supplements. The Eurolysine Company in Amiens, France, recently invested $27 million to double its production of lysine. There has been an increase in the use of the amino acids glycine and alanine as flavoring agents and in that of cysteine as a bread texturizer. It has also been suggested that gastric ulcers be treated with the amino acids glutamine and histidine and that liver disorders be treated with arginine.

How will the fermentation industry respond to the higher demand? One can expect the commercial production of more amino acids and an improvement in yield brought about by the application of recombinant-DNA techniques to the circumvention of intracellular regulatory mechanisms. Where the regulatory mechanisms are too complex to be circumvented conveniently, a precursor of the amino acid can be made by fermentation and subsequently transformed enzymatically.

Perhaps the most significant contribution of genetic engineering to amino acid synthesis will be to make possible the fermentative production of methionine, of which 105,000 tons per year are chemically synthesized. The biological synthesis has always been a goal, but microorganisms that yield large amounts of methionine have not been obtained by mutation and selection. The approach has failed because it can exploit only existing biochemical pathways. With recombinant-DNA techniques one should be able to introduce the necessary new pathways and regulatory mechanisms.

An alternative to the commercial production of methionine would be the production of a protein rich in methionine. When methionine is added directly to animal feed, the feed is bitter. If the bitterness could be eliminated, methionine might be added to human food as a nutritional supplement. A protein incorporating large amounts of methionine is more palatable than methionine alone.

It is only a matter of time before the synthetic-chemicals industry cedes to biological methods the production of all amino acids. The new genetic-programming techniques may render obsolete the "Organic Chemist's Ode":

> Lord, I fall upon my knees
> And pray that all my syntheses
> May no longer be inferior
> To those conducted by bacteria.

7

Production Methods in
Industrial Microbiology

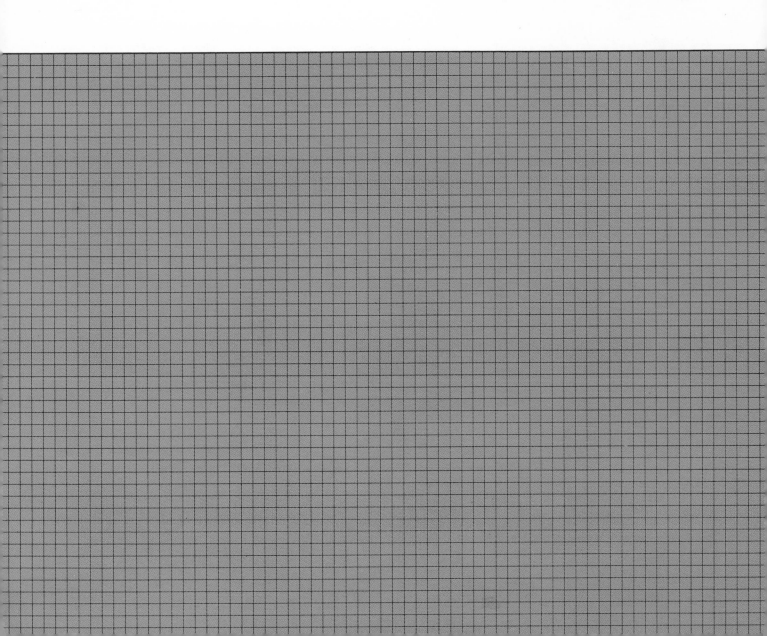

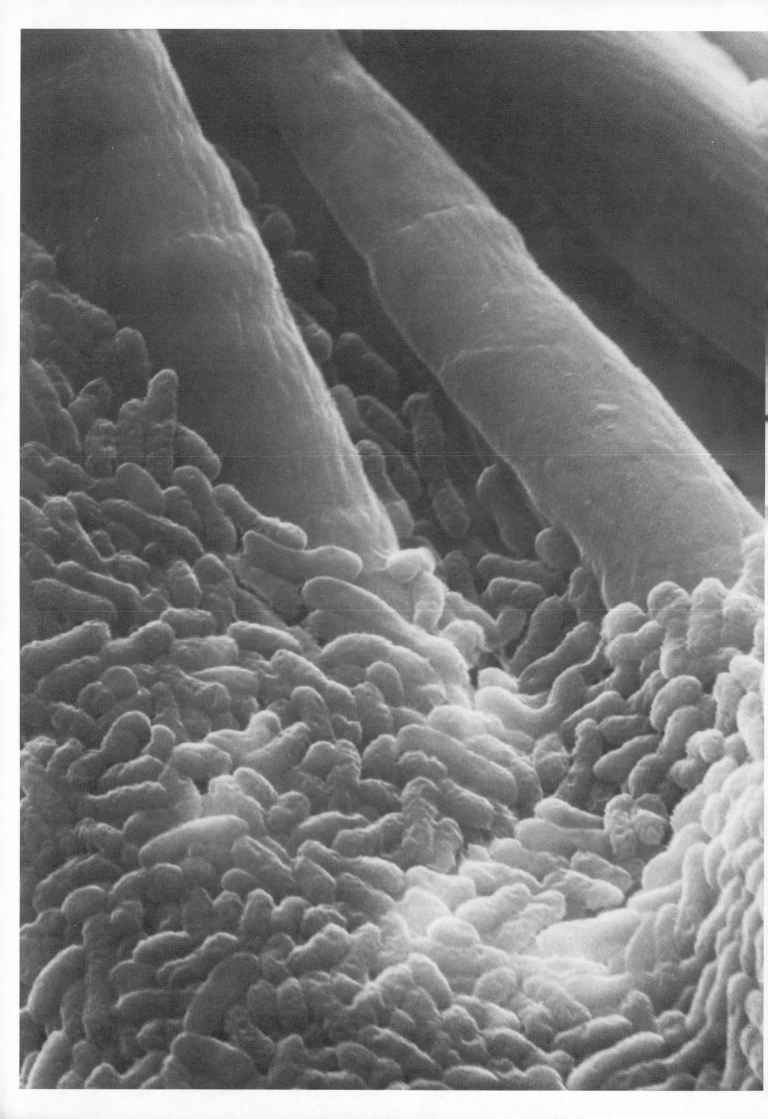

Production Methods in Industrial Microbiology

BY ELMER L. GADEN, JR.

Traditional practice combines with the scale on which most products are made to favor manufacture in batches. Newer, continuous methods, however, are being explored

In the applications of microbiology to industry the most distinctive element is usually the biological one: the exploitation of a living organism for the manufacture of a useful substance. As is set forth elsewhere in this *Scientific American* book, the methods of genetic engineering promise to increase the efficiency and the versatility of the organisms on which such industries depend. It must be kept in mind, however, that a biological process can attain its full utility only when it is adapted to a context of production. Raw materials must be brought together with living cells or with components (notably enzymes) extracted from the cells; conditions that favor the biochemical transformation of the raw materials into products must be maintained; often a product must be isolated from other substances with which it is mixed. Hence industrial microbiology requires not only microorganisms but also an environment in which the organisms can grow and a technology for handling them and their products. Both the environment and the technology are generally provided by a system of vats, pipes, pumps, valves and other devices. It follows that genetic engineering is only one factor in the success of a biological industry; the contributions of process engineering are also essential.

For any given biochemical procedure there are many ways to organize a plant of industrial scale. So far, however, only a few ways are practiced; they can be divided into two broad categories: batch processes and continuous processes. In a batch process a vessel is filled with starting materials, often including the microorganisms themselves. The biochemical conversion takes place in the vessel over a period that can range from a few hours to several days. Ultimately the vessel is emptied, the product is purified and a new batch is started. In a continuous process the raw materials are supplied and the finished products are withdrawn in a steady stream. With such a process all stages in the biochemical conversion must proceed simultaneously and at essentially the same rate. The batch process can be likened to the operation of a steel mill, whereas the continuous process stands in closer analogy to the operation of a petroleum refinery.

The choice between the batch and the continuous methods must be made on economic grounds. In general the continuous methods are best suited for a large volume of production; nevertheless, up to now most products of industrial microbiology have been made in batches. The reasons, which I shall take up below, have to do in part with the biological nature of the processes and in part with the scale on which most industrial microbiology is conducted. They may continue to favor batch methods for some time to come.

As the preceding articles have shown, the industrial processes carried out by microbiological means vary greatly in their details. In broadest outline, however, they are much the same. From the point of view of the technologist the biological steps can almost always be understood in terms of the chemical process of catalysis. The transformation of a substrate into the desired product is accelerated by the presence of a catalyst and is thereby selectively favored over other possible reactions.

According to this scheme, a microorganism is merely a catalyst of exceptional complexity. For example, the yeast employed in making beer or wine can be regarded as a catalyst for the conversion of sugars into ethanol and carbon dioxide. Of course, the actual agents of chemical change are the enzymes made by the organism, and in some instances the enzyme itself can serve in place of the complete cell. In the brewing industry this practice is well established: an enzyme separated from barley malt or from a mold breaks down starch into molecules of sugar.

More commonly, however, the biological transformation of the substrate includes several interlocking chemical reactions, each reaction catalyzed by a separate enzyme. Where the biological process is the synthesis of a complex molecule, such as an antibiotic, or of a protein, such as insulin, entire systems of enzymes are recruited to the task. Such systems have not yet been made to function outside the living cell. Indeed, where the product is the cell itself, as in the culturing of baker's yeast, all the enzymes that participate in the metabolism of the cell can be considered components of the catalytic system.

A distinguishing feature of a biological catalyst, and a feature that has a ma-

BACTERIA ARE IMMOBILIZED on cotton fibers for the manufacture of an industrial alcohol (ethanol) in a scanning electron micrograph made by Carl E. Shively of Alfred University. The bacteria, of the species *Zymomonas mobilis*, have been employed for centuries in Central America for making fermented beverages. One such beverage is pulque, made by fermenting the juice of the agave plant. It now appears the bacteria are more efficient than yeast at converting carbohydrates into ethanol. To make the micrograph, cotton fibers were woven onto a supporting plastic mesh and inoculated with the bacteria. The mesh was fitted into a horizontal glass chamber 22 inches long. Nutrients including glucose were introduced at one end; the spent medium, including ethanol, flowed out at the other end. After 15 days of operation a sample of the immobilized bacteria was taken. It remains uncertain whether the bacteria are entangled among the fibers or are held to them by a force such as electrostatic attraction.

84

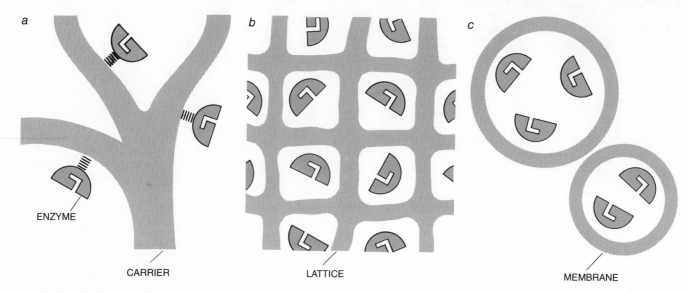

ENZYME IS IMMOBILIZED for long-term employment as a catalyst in a reactor vessel by any of several methods. The enzyme molecules can be held by adsorption or by chemical bonding to a solid carrier such as cellulose (*a*), they can be trapped in the lattice formed by a permeable polymer such as a silica gel (*b*), or they can be trapped in spherical capsules made of semipermeable polymer membrane (*c*).

jor influence on the design of an industrial plant, is the need of the catalyst for a precisely controlled milieu. Even an inorganic catalyst operates best at some particular combination of temperature, pressure and other physical conditions. The constraints on the functioning of a biological catalyst are much more stringent. The temperature and the *p*H cannot be allowed to vary beyond a narrow range. Moreover, when the biological catalyst consists of living cells, the medium in which the reaction takes place must furnish all the nutrients and other substances needed to sustain growth.

The medium serves as a reservoir for the substrate and the nutrients and it provides the environment where the substrate and the catalyst interact. The dominant component of the medium is almost always water. Even where microorganisms grow on a solid substrate, such as grain or hay, the substrate must be dampened in order to support microbial or enzymatic action. Although some microorganisms and enzymes can be preserved by careful drying, they have no catalytic activity in the absence of water.

In addition to providing a suitable aqueous environment, the medium must meet the nutritional needs of the microorganism. A primary need is a source of carbon, which ordinarily supplies the energy for metabolism. In some cases the carbon source is also the substrate of the catalyzed reaction, as in the fermentation of sugar to yield ethanol. The commonest sources of carbon are the carbohydrates, such as starch and sugar. In the 1960's, however, certain hydrocarbons from petroleum and some natural fats such as soybean oil were considered as alternative sources

of carbon and energy. Many microorganisms of industrial importance can exist on such materials, although sometimes a period of adaptation is needed. Alternative substrates were of interest then because grain was expensive and petroleum was comparatively cheap. The price structure has clearly changed; indeed, the biological conversion of carbohydrates into hydrocarbon fuels is now being considered. Nevertheless, there are a few applications in which petroleum fractions poorly suited to the making of gasoline serve as ingredients in a biological process.

The consideration given to hydrocarbon substrates is an apt illustration of the versatility of microorganisms; it should be pointed out that the technology supporting the microorganisms cannot always adapt to change as readily. Hydrocarbons and fats incorporate less oxygen than carbohydrates do, and so more oxygen must be supplied. Up to three times as much oxygen may be needed, and the heat released when the substrate is consumed is greater by a similar ratio. Occasionally the equipment available has been unable to provide sufficient cooling when petroleum or fats are introduced as raw materials.

After carbon is provided, the nutrients needed in substantial quantities are sources of nitrogen and phosphorus. Both elements are incorporated into the structural and functional molecules of the cell. They also become part of the product molecules. A number of other nutrients, such as vitamins and metal ions, are required in smaller amounts. Again, some of these "micronutrients" appear as part of the product molecules. For example, in the manufacture of cobalamin, or vitamin B_{12}, a supply of cobalt must be ensured because each

molecule of the vitamin incorporates an atom of cobalt.

Oxygen is another element whose supply must be taken into consideration. Some fermentative organisms are strictly anaerobic, and so oxygen must be excluded from their environment. Where oxygen is needed for metabolism, however, the need is absolute. Filtered air is the usual source of supply, but with the recent increases in the price of electricity the cost of pumping large volumes of air has become significant. The cryogenic fractionation of air into its component gases offers a possible remedy. By employing enriched air, which has more oxygen than the usual 21 percent, the volume to be pumped can be reduced.

Whatever the chemical composition of the medium, it is imperative that all the components be thoroughly mixed, so that the microorganism has ready access to the available nutrients and to the substrate. Most bacteria and some yeasts commonly grow as individual cells or as aggregates of a few cells each, and they remain suspended in the medium. Even in a dense population they have little effect on the physical properties of the fluid in which they are growing, apart from making it cloudy. In some cases, however, the cells secrete natural polymers that greatly increase the viscosity of the medium; they can also form large aggregates or grow as a slimy film on a surface.

Other bacteria and yeasts and most molds have a quite different growth habit. When they are allowed to grow undisturbed, they form a tough, continuous film, and when they are dispersed throughout a fluid medium by vigorous stirring, they create a fibrous pulp. If enough nutrients can be supplied, the cells proliferate until the suspension has the consistency of oatmeal. Such changes in the medium have an effect on process technology. For example, oxygen bubbled through a watery medium is readily absorbed and transported to the sites where it is needed. In a pulpy or gelatinous medium, on the other hand, the absorption and transport of oxygen are impeded.

It bears emphasizing that the biological steps in an industrial process are seldom the only steps. The pretreatment of raw materials and the extraction, purification and further alteration of products are major factors in the economics of industrial microbiology. The importance of the nonbiological stages can be made plain by considering two examples: the production of ethanol and that of cobalamin.

The commonest raw materials for the production of ethanol are molasses, which is about 50 percent sugar, and corn, in which the major carbohydrate is starch; yeasts can metabolize the sugar but not the starch. Either material, however, requires considerable preparation before the yeast cells can be introduced. Molasses must be diluted and made more acidic; it may also be necessary to add minor nutrients and to remove other substances (such as iron) that are sometimes present in concentrations high enough to inhibit the growth of the yeast or the formation of the alcohol. When corn is the raw material, the grain is cooked to make the starch soluble; then the starch must be converted into sugar by the action of enzymes from malt. As with molasses, nutrients may have to be added and the *p*H may have to be adjusted. All these procedures require time, equipment and energy.

When fermentation of the sugar is complete, ethanol makes up from 6 to 8 percent of the spent medium, which also includes by-products, wastes, unconsumed nutrients and many minor constituents. The ethanol is recovered and purified by distillation. In the fermentation of grain the solid residue is also of value; it is recovered by evaporation and drying. The residue consists of dead yeast cells, grain proteins and other materials, and it makes a nutritious animal feed. The sale of the residue contributes to the economic feasibility of making ethanol from grain.

In the manufacture of cobalamin and related substances the biological catalyst is not a yeast but a bacterium; several species can carry out the synthesis. The preparation of the starter culture of bacteria and of the growth medium are much the same as they are for yeast, although more stringent controls are needed to avoid contamination of the culture. The key difference is encountered when the conversion is completed: most of the vitamin is not excreted by

the bacteria, as ethanol is by yeasts, but is retained within the cells. The cells must therefore be treated in a way that will release the cobalamin and the related substances. It is then possible to extract a crude product of roughly 80 percent purity that can serve as a vitamin supplement in animal feed. The purity of from 95 to 98 percent required for medicinal purposes can be attained only through a much more complex and thoroughgoing extraction procedure.

A concern common to almost all biological technologies is the need to maintain aseptic conditions. The reason is that most products of such technologies are synthesized by a pure culture: a population of organisms made up of a single species or even a single strain of a species. If foreign organisms contaminate the culture, they can disrupt its operation in several ways. They can directly inhibit or interfere with the biological catalyst, whether it is an isolated enzyme or a living cell; they may even destroy the catalyst entirely. Alternatively, the contaminating organisms may leave the catalyst unaffected but destroy the product. Further, the foreign organisms can introduce noxious substances that are difficult to separate from the product. In the manufacture of pharmaceuticals the risk of toxic impurities is of particular concern.

In order to avoid contamination all materials entering the culture medium are sterilized, including the large volumes of air required for aerobic processes. Foreign organisms are filtered from the air by a deep bed of glass wool, which can itself be sterilized at intervals with steam. Steam is also employed to sterilize reactor vessels, pipelines and other surfaces with which the medium comes in contact. The apparatus must be designed and operated so that the opportunities for invasion by unwanted organisms are minimized. Maintaining the integrity of various entry and exit points in the system is notably difficult. In spite

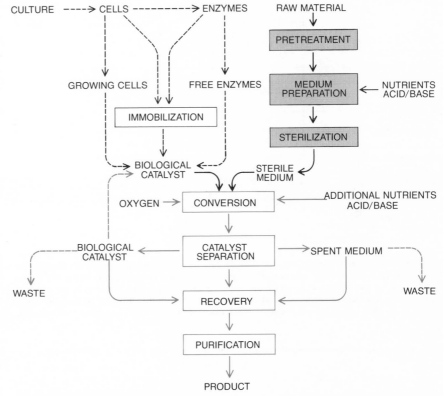

SEQUENCE OF STEPS in the industrial application of a bacterium, a yeast or a mold as a biological catalyst varies from one process to another, but it always follows the outline shown. Solid lines represent the main steps common to all processes; broken lines represent options. The preparation of the catalyst is shown at the upper left. Whole cells are often employed; increasingly, however, enzymes (the true agents of change) are isolated from the cells. Increasingly too the cells or the enzymes are immobilized to trap them in the reactor vessel. The preparation of the medium is shown at the upper right. Typically the medium is aqueous and carries in solution or suspension the substrates: the substances the catalyst transforms. Where the catalyst consists of living cells the medium must also supply nutrients. The *p*H of the medium is adjusted and the medium is sterilized in an effort to prevent contamination by foreign organisms. The synthesis and subsequent isolation of the product are shown at the bottom. First the catalyst acts on its substrates; then the catalyst and the medium are separated. Some products (such as vitamin B$_{12}$) remain bound to the catalyst. Others (such as penicillin) are excreted into the medium. The product emerges in a dilute solution from which it must be purified.

NUTRIENT	RAW MATERIAL	PRETREATMENT
CARBON SOURCE	CORN SUGAR	
GLUCOSE	MOLASSES	"INVERSION": SUCROSE → GLUCOSE + FRUCTOSE
	STARCH	COOKING FOLLOWED BY SACCHARIFICATION: STARCH → GLUCOSE
	CELLULOSE	GRINDING AND COOKING FOLLOWED BY SACCHARIFICATION
FATS	VEGETABLE OILS	
HYDROCARBONS	PETROLEUM FRACTIONS	PURIFICATION BY DISTILLATION
NITROGEN SOURCE	SOYBEAN MEAL	
PROTEIN	CORNSTEEP LIQUOR (FROM CORN MILLING)	
	DISTILLERS' SOLUBLES (FROM ALCOHOLIC-BEVERAGE MANUFACTURE)	
AMMONIA	PURE AMMONIA OR ITS CHEMICAL COMPOUNDS	
NITRATE	NITRATE SALTS	
NITROGEN	AIR (FOR NITROGEN-FIXING ORGANISMS)	
PHOSPHORUS SOURCE	PHOSPHATE SALTS	

NUTRIENTS for the biological catalyst include sources of carbon, nitrogen and phosphorus. The choice of a source is made on economic as well as biological grounds; as a source of carbon, for example, carbohydrates from grain and other plant products are the ones in widest use. Many sources require a specific pretreatment. Starch, for example, must be cooked and then broken down into sugar (glucose) before most microorganisms can convert it into ethanol.

of all precautions, the potential for human error or mechanical failure is great, and serious losses are not uncommon.

In a batch process most or all of the constituents of the medium are combined with the biological catalyst at the start. Typically they are mixed in a cylindrical vessel whose height is from 2.5 to four times its diameter. The capacity of the vessel ranges from a few hundred gallons to several tens of thousands of gallons. In some applications the volume may be still greater. Before about 1950, when industrial alcohols such as butanol were made by fermentation, the process was done in spherical tanks with a capacity of up to 500,000 gallons.

When the vessel is intended for the manufacture of products such as antibiotics pure enough for pharmaceutical use, it is constructed of stainless steel or of an alloy of comparable inertness. For less stringent applications a vessel of carbon steel or of steel with a coating resistant to corrosion will suffice.

After the vessel is sterilized the starting materials enter it by means of a number of tubes and pipes. Steam lines bathe the various entry points, attesting to the requirement that the process operate aseptically. In the vessel the catalyst and the constituents of the medium are mixed by a rotating central shaft that carries several impellers. Coils inside the vessel or jackets that surround

it provide heating for sterilization and either heating or cooling to maintain an optimum operating temperature. Equipment to monitor and control the temperature and the pH of the medium is common. Somewhat less often one finds equipment to monitor and control the concentration of oxygen dissolved in the medium.

As the biological conversion proceeds, nutrients may be added to the mixture to sustain the growth of the organisms; if the process is an aerobic one, oxygen must be supplied continuously. Meanwhile samples of the mixture and of gaseous by-products can be removed for analysis by means of other tubes and pipes. When the concentration of the product reaches its maximum level, the finished batch is removed by means of still another pipe.

The monitoring of conditions in the reaction vessel during the progress of a batch is a matter of urgent concern. As mentioned above, the temperature, the pH and the concentration of dissolved oxygen can be recorded continuously. It is also useful to know the concentration of the substances that serve as sources of carbon, nitrogen and phosphorus, and perhaps also the concentration of a critical micronutrient. Of still greater interest are the amount of biological catalyst present and its level of activity. With the methods now available it is not possible to determine these values directly in the reaction vessel. Instead samples must be withdrawn for laboratory assay.

In principle a continuous industrial process often has numerous advantages over one done in sequential batches. For example, the continuous process usually has the potential of a higher volume of production for an installation of a given size. One approach to continuous operation is simply to modify a batch reactor so that fresh nutrient and substrate can continually be added and the products of the reaction can continually be removed. A device that has been modified in this way is called a continuous stirred-tank reactor. It can be controlled in two basic ways. In the first method the turbidity, or cloudiness, of the outlet stream is monitored. The turbidity, which is caused by microbial growth, yields a measure of the rate at which cells leave the tank. The measure controls the rate at which fresh nutrient is admitted. The reactor is called a turbidostat.

The second method of controlling a continuous stirred-tank reactor is simpler and can be applied in cases where the product of the reaction does not consist of cells. The reactor is called a chemostat, and it controls the reaction by monitoring not the output stream but the input stream. In chemostatic operation the concentration of a critical nutrient in the feed to the reactor is fixed at a level such that the other nutrients are abundant. The level of the critical nutrient then limits the extent to which the microorganisms can proliferate. A drawback of the method is that the stream leaving the reactor includes appreciable amounts of nutrients that are not consumed.

Both mathematical models of the action of a chemostat and experimental studies show that a single-stage chemostat of practical size cannot yield a high concentration of a product and low concentrations of unconsumed raw materials. The inefficiency is particularly great in the synthesis of products such as penicillin, which is a secondary metabolite: it is made by living cells but its synthesis does not arise in the course of the metabolism that keeps the cells alive and growing. It is characteristic of the industrial production of a secondary metabolite that the proliferation of the organism precedes the accumulation of the metabolite by a significant margin. It is also characteristic that the conditions of temperature, *p*H and so on that are optimum for the growth of the organism differ from those that are optimum for the formation of the product. A single-stage chemostat can offer at best a compromise between the two conflicting optimum environments. Conditions more favorable to each stage in the life cycle might be provided by devising a chemostat in which the fluid stream cascades through a series of tanks. Such a device would be difficult to operate, however, and it seems unlikely that the volume of production would be large enough to justify the investment. The one application in which single-stage and multistage chemostats have found use is the biological treatment of wastes.

Another potential benefit of a continuous method of operation is in reducing losses of catalyst. A definitive characteristic of a catalyst is that it is not consumed in the course of a reaction; in most batch processes, however, the catalyst is discarded with the spent medium. The waste can be costly: the value of a biological catalyst is at least equal to the value of the nutrients consumed in growing the cells.

There are two ways to limit the amount of catalyst that is lost from the cycle. One way is recycling. In modern industrial applications, however, it is difficult to take living cells from a stream of fluid and return them to the reactor. Often the cells are damaged, and still more often the reactor becomes contaminated with foreign organisms.

The other way to reduce losses is to keep the catalyst inside the reactor. The

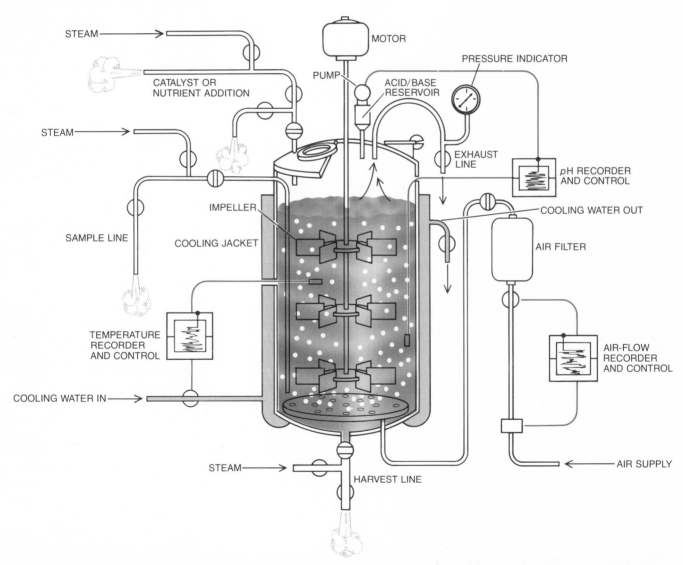

BATCH REACTOR is employed for most current applications of industrial microbiology. In essence the reactor is a vessel in which quantities of the medium and the biological catalyst are mixed and then given an optimum environment in which to react. The temperature and the *p*H are regulated. Filtered air, sometimes enriched with oxygen, is bubbled through the mixture. Samples are removed for chemical and biological assay. Two strategies are employed to prevent contamination: steam is directed through the various inlets to keep them sterilized, and the pressure inside the vessel is maintained at a value greater than atmospheric pressure. At the end of a period that can range from hours to days the batch is drained from the vessel so that the product of the reaction can be isolated and purified.

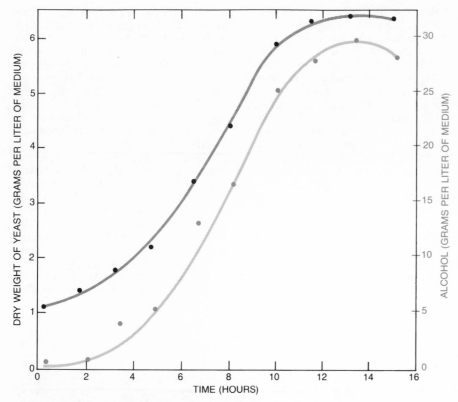

PRIMARY METABOLITE is synthesized by a microorganism in the course of the metabolic processes that keep the cells alive and growing. In a reactor vessel a primary metabolite accumulates in tandem with the accumulation of the cells that synthesize it. The graph shows the accumulation of yeast cells (*black*) and the concomitant accumulation of ethanol (*color*).

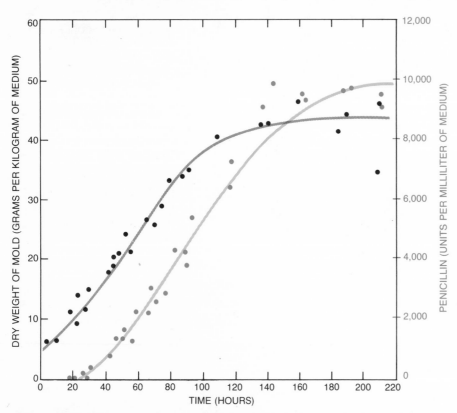

SECONDARY METABOLITE is not formed as a direct result of the metabolism that keeps the cells alive. Hence the accumulation of a secondary metabolite in a reactor vessel lags behind the growth of the cells that produce it. The graph shows the accumulation of mold cells (*black*) and the subsequent accumulation of penicillin (*color*). The values of temperature and pH that are best for the growth of cells are seldom best for the synthesis of a secondary metabolite. In a batch process one seeks a compromise between the two sets of optimum conditions.

commonest technique employs a packed bed: a solid support on which the cells are encouraged to grow. A continuous reactor of this kind has long been employed for making vinegar. Diluted wine or fermented cider is percolated through a bed on which a culture of microorganisms that oxidize ethanol into acetic acid has been established. The microorganisms, which constitute a mixed culture rather than a pure one, form a slimy film on the surface of the bed. A conceptually similar scheme has evolved for the treatment of sewage and other wastes. The waste stream trickles through a filter of bits of stone, ceramic or plastic, where a microbial film traps the waste particles and oxidizes them.

In recent years several new methods of immobilizing both enzymes and whole cells have been devised. The earliest and simplest of the methods is adsorption: the enzyme molecules or the cells adhere loosely (without chemical bonding) to the surface of a material such as alumina, charcoal, clay or cellulose. Eventually the adsorbed agent washes away, but surprisingly long useful lives have been reported. For an isolated enzyme a firmer attachment can be created by forming a chemical bond between the enzyme molecule and a support material, which might be cellulose, glass or a manmade polymer. The result is a stable preparation capable of extended service; moreover, the fixation of the enzyme apparently interferes little with its activity. In both of these techniques the usual practice has been to divide the support material into small particles, creating a packed bed. It is also possible to bind the catalytic agent to a continuous, flat membrane or to the inner surface of a tube.

A third method of immobilization, applicable both to cells and to enzymes, is entrapment in a polymer matrix. When starch, a silica gel or certain other polymers are permeated with water, they form a meshwork of fibers with voids where enzyme molecules or cells can become trapped. A limitation of the technique is that molecules of the substrate, of the product and of the nutrients must diffuse through the solid matrix, reducing the rate of reaction. There are compensating advantages, notably that living cells can be held firmly in place without damage.

In the technique called microencapsulation enzymes or cells are enclosed in a spherical polymer membrane. The resulting capsules range in diameter from five to 300 micrometers; they look much like enlarged cells. The composition of the membrane is chosen so that it is semipermeable: the comparatively small molecules of the substrate and of the product pass through the membrane freely, but the larger molecules of an enzyme or the still larger structure of a cell cannot escape.

The newest approach to the retention of the catalyst is copied from the chemical industry. It is the fluid-bed reactor. The basic mechanism of such a reactor is a vertical tube that widens toward the top. The input stream is forced upward from the bottom, and so as the cross-sectional area increases with height the velocity of the fluid decreases. The catalyst, which is suspended in the fluid stream, settles at a level in the reactor below the level where the liquid stream is removed. The catalyst must of course have a form that keeps it suspended in the reactor. In some applications the catalyst (which can be either a microorganism or an enzyme) is immobilized on particles of coal.

Given the evident advantages of the

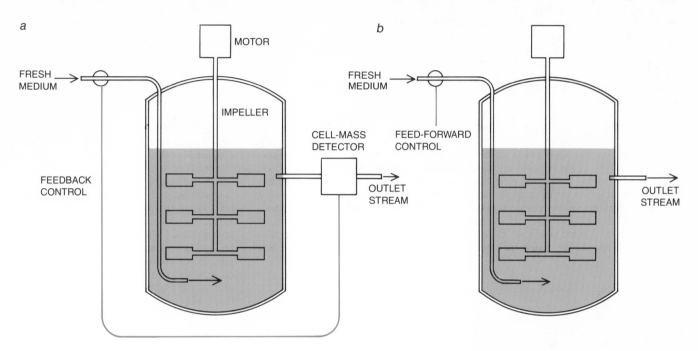

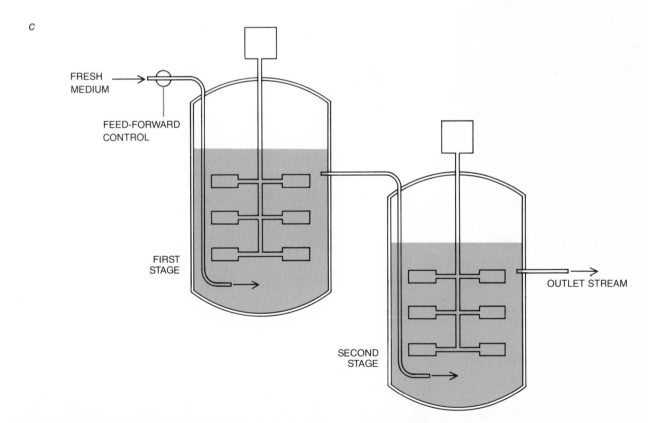

CONTINUOUS STIRRED-TANK REACTORS represent efforts to adapt batch technology to continuous operation. In a turbidostat (*a*) the rate at which cells leave the reactor vessel (as measured by the turbidity, or cloudiness, of the outlet stream) governs the rate at which fresh nutrients enter. In a chemostat (*b*) the rate at which a critical **nutrient enters the reactor vessel is adjusted so that it limits the rate of the reaction. In a two-stage chemostat (*c*) the mechanism of control is the same but conditions in the two vessels can differ. Such an arrangement is useful, for example, in the production of a secondary metabolite or in the successive stages in the treatment of wastes.**

continuous-stream method of operation, why has it made little progress in displacing batch methods? Some of the impediments are strictly technical. For example, it is harder to maintain aseptic conditions in a continuous reactor. When a product is made in batches, all the components of the apparatus can be sterilized after each batch, so that any contaminating organisms have only a limited period available for growth and proliferation. If the economies of continuous operation are to be fully achieved, the reactor must operate without interruption for long periods. Any microorganism that breached the barriers to contamination would grow unchecked.

Difficulties of this kind could probably be overcome if there were enough economic incentive to do so. Actually the volume of production in most biological processes remains comparatively small, so that the efficiency of a continuous product stream could not be fully exploited. Moreover, batch methods

offer great operational flexibility: a reaction vessel and the associated apparatus can produce a single batch of one product and can then be turned immediately to the production of a product in greater demand. This versatility is particularly important in the pharmaceutical industry, where the variety of products is large but the quantity of each product is small. It is notable that the one area where continuous processes are predominant, namely the treatment of sewage, is by far the largest microbiological industry in terms of the volume of material processed.

When the biological conversion is completed, the product or products must be separated from the spent medium and the product must be purified. Here a number of difficulties arise that are peculiar to industrial microbiology. First, many products are chemically fragile. It may be necessary to control carefully the temperature and the pH of the mixture. It may also prove necessary

to exclude even traces of metals or other impurities. Second, the product is usually dissolved or suspended in a large quantity of water. Either the water must be removed from the product or the product must be removed from the water. Evaporation or distillation is sometimes effective. Distillation, however, is energy-intensive; its cost can constitute a substantial debit against the value of the product. Moreover, when the product molecule is fragile, evaporation or distillation would destroy it. A number of less stressful techniques have therefore been developed.

One technique is solvent extraction. The aqueous solution bearing the product is combined with a second liquid, immiscible with water, in which the product has a greater solubility. Another technique is adsorption. Here the product molecules leave the solution when they become attached to the surface of a solid material. Membrane separations, where the liquid is driven through a membrane that blocks the passage of the product, are finding increased application. Most of the techniques are applied to a stream of liquid from which the biological catalyst has been removed by a method such as filtration or separation in a centrifuge. Finally the product is purified. The volume of material that must be processed at this point is generally small and the techniques are specific to the product.

The exploitation of biochemical processes began with food and drink, and when industrial applications emerged in the 19th century, they were rooted in the sequence of steps that had become established through traditional practices. The first step remained the purification of raw materials and the development of a seed, or starter, culture: a natural population of microorganisms in which the organism whose catalytic activity is useful is dominant. Then came the biological conversion itself. With the manufacture of ethanol for use as a solvent rather than as a drink the need arose for techniques by which specific products could be recovered and purified.

The persistence of the traditional methods cannot be attributed simply to the inflexibility of the early industrial practitioners. The workings of microorganisms in an industrial process are complex and delicate, and many aspects of their functioning are still daunting today, when a large and prosperous industry operates throughout the world. For the most part the industry is limited to the small-scale batch production of substances of great value. The world economy is changing, however. If industrial microbiology should become competitive for the manufacture of products such as fuels and industrial chemicals, a body of knowledge and experience will be available to guide the effort.

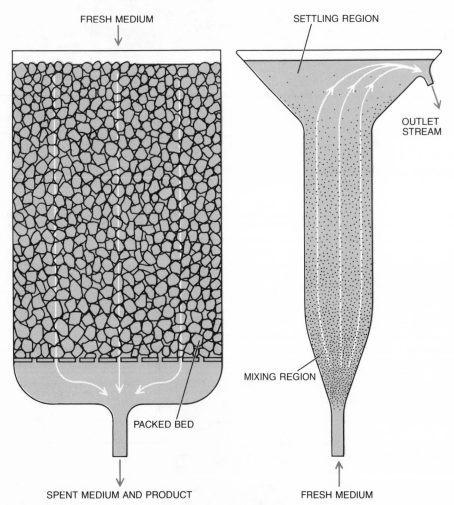

PACKED BED is a well-tried technology in industries such as the treatment of wastes and the manufacture of vinegar. The catalyst is a slimy film of microorganisms that adheres to a solid bed. Often it is a natural culture. The medium trickles through the bed from above.

FLUID BED is new to industrial microbiology. The biological catalyst is immobilized on particles that are suspended in an upwelling stream of fresh fluid medium. A widening at the top of the reactor vessel slows the fluid and thus keeps the catalyst inside the vessel.

8

Agricultural Microbiology

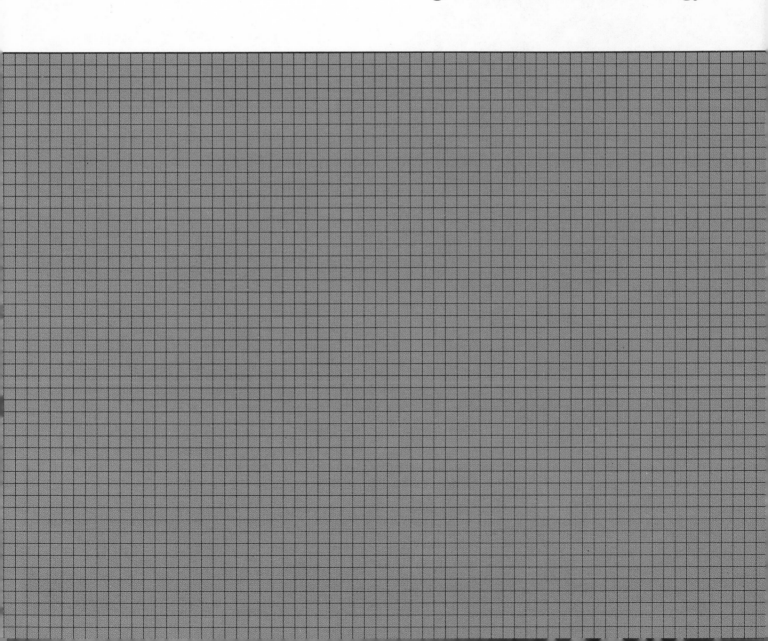

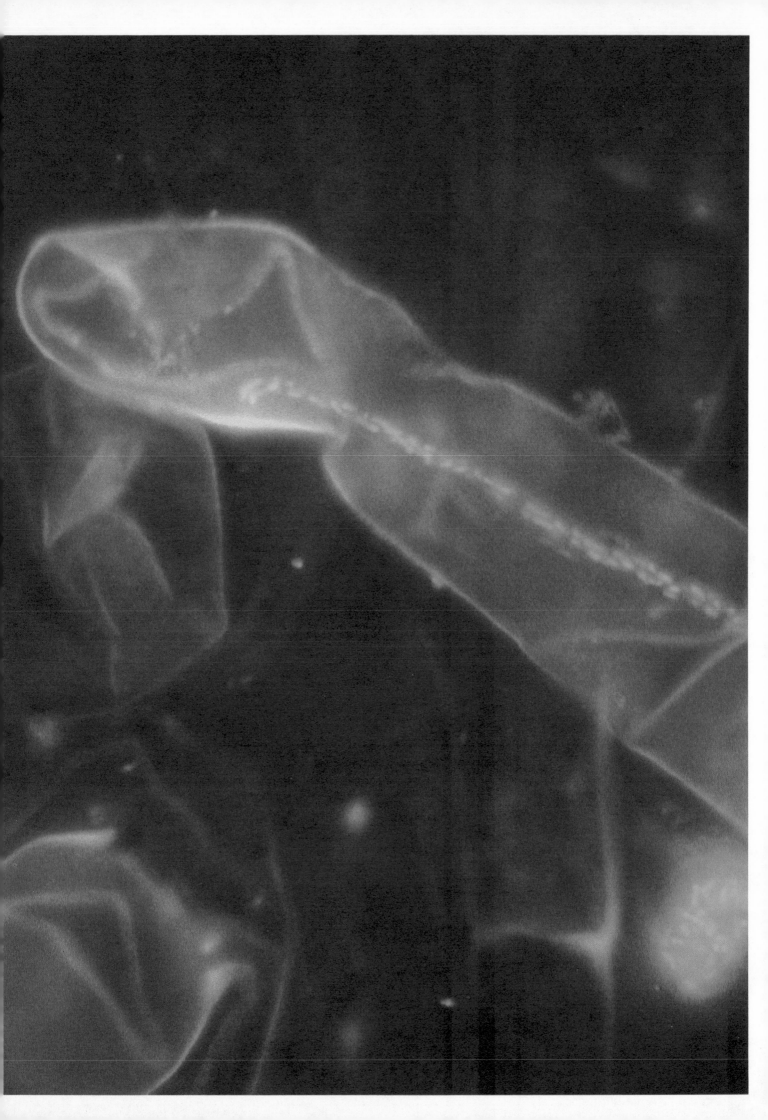

Agricultural Microbiology

BY WINSTON J. BRILL

Introducing new genes into crop plants by recombinant-DNA methods is difficult and not in immediate prospect. Much progress can be made, however, by manipulating the microorganisms that live with plants

A handful of soil scooped from the fields of a mechanized farm is the locus of an unruly turmoil of competing microorganisms. In the soil thousands of strains contend for nutrients and energy, in the process altering the chemistry of the soil with the products of their metabolism. Moreover, the microorganisms themselves evolve in response to stresses imposed by their environment, including the stresses induced by the evolution of their fellow species. Suppose into the teeming genetic marketplace a new colony of bacteria is introduced, one selected, say, for the ability to invade the roots of certain crop plants. Even if the new bacteria can survive the competition and adapt to the changeable environment, they may not be able to carry out their intended function. They may find that the root system is already occupied by other strains of microorganisms.

Interference of this kind is commonplace in agricultural practice; it serves to illustrate the difference between the open field and the fermentation tank. There are also similarities: in both agriculture and industrial microbiology the aim is to meet human needs by the selective breeding, culturing and harvesting of living organisms. If the fermentation tank can be taken to represent an agricultural environment, however, it is one subject to exceptionally precise control. For example, the population of microorganisms is usually limited to a single species. In considering the potential applications of microbiology to agriculture, one must begin to explore the subtle question of the interaction of microorganisms with one another and with the biosphere as a whole.

The growing demand for food and other agricultural products is ample practical justification for undertaking the enormous research effort that will be needed in order to apply the methods of microbiology to agriculture. It has been suggested that the "engineering" of crop plants and of the soil microorganisms on which they depend may yield hybrid grains capable of obtaining their supply of nitrogen directly from the atmosphere. No crop plant now has this capacity; the nitrogen must be fixed, or converted into a biologically useful form, either by microorganisms or by the industrial manufacture of nitrogenous fertilizers, a process that calls for a large expenditure of fossil fuel. American farmers spend about $1 billion a year on nitrogenous fertilizers for the corn crop alone, and so the program of research on nitrogen fixation is a growing one.

Other lines of biological investigation may lead to the acceleration of photosynthesis and to the development of crops that can be grown on saline or highly acidic soil. These are ambitious goals, unlikely to be achieved commercially in less than 10 years. Nevertheless, a great deal of rational exploitation and modification of the microbiological environment of the farm is within reach during the next decade. The agricultural community has only recently begun to recognize the potential of microbiological techniques in plant-cell research. Recombinant-DNA technology may lead to improvements in existing crops and to the development of entirely new crop types. So far, however, its most important effect has probably been on industrial laboratories, which have been alerted to the possibility of applying biological methods to agriculture. Work of this kind is now under way in several dozen such establishments, focusing primarily on the design of microorganisms important to agriculture or on the application of microbiological techniques to the manipulation of plants. Perhaps the most significant indicator of the revolutionary nature of the work is that almost all these laboratories have been established within the past two years.

How can microbiological techniques be applied to traditional agricultural practices? There are three main strategies. First, microorganisms found to be beneficial to plants (or designed for this purpose) can be bred and grown in fermentation tanks for later introduction into the soil. Second, individual cells can be isolated from plant tissue and grown in nutrient solution. In such cell cultures the rate of mutation can be increased, making possible the selection of promising strains, the development of hybrid strains that would not be obtainable by standard breeding techniques and the manufacture in large fermenters of certain plant products, such as digitalis, pyrethrins (which are natural pesticides) and licorice.

Third, foreign genetic material can in some cases be introduced into plant cells, a practice that could open the way for direct genetic engineering of the plants themselves. Such techniques are still primitive when they are compared with the achievements of recombinant-DNA technology with bacteria. Indeed, the reverse strategy of inserting plant genes into bacteria, so that plant proteins can be made by culturing bacteria in standard fermentation tanks, may

MICROORGANISMS ESSENTIAL TO AGRICULTURE include bacteria of the genus *Rhizobium*, which are seen infecting a root hair of a clover plant. In the photomicrograph, which was made by B. Ben Bohlool of the University of Hawaii at Manoa, the bacteria have been labeled with a fluorescent dye; they appear as bright chartreuse flecks. They form an infection thread that extends from the tip of the root hair into the interior of the root. *Rhizobium* bacteria live symbiotically in the roots of clover and other legumes; the bacteria supply the plant with fixed nitrogen necessary to photosynthesis and the plant nourishes the bacteria. Such interactions of microorganisms with plants suggest several ways the new methods of genetic engineering might contribute to agricultural practice. For example, the nitrogen-fixing capacity of the bacteria might be increased by genetic modification, or the bacteria might be induced to colonize plants other than legumes. If the modified organisms are to be effective, however, they must be able to compete successfully with strains indigenous to the open field.

well precede the genetic engineering of plant DNA.

The exploitation of microorganisms in the soil is hardly new to agriculture. It was well known in Roman times that legumes such as beans, peanuts, alfalfa, soybeans, peas, clover and lupine enhanced the fertility of the soil. Soil from fields where legumes had been grown was added to fields that were to be assigned to legume cultivation for the first time. The Romans could not have

known that the underlying justification of their sound empirical practice is the presence of bacteria of the genus *Rhizobium*, which infect the roots of some legumes and fix atmospheric nitrogen [see "Biological Nitrogen Fixation," by Winston J. Brill; Scientific American, Offprint 922]. As the term infect suggests, the introduction of *Rhizobium* into legumes resembles a disease process, but it is one in which the plant cooperates. The *Rhizobium* bacteria enter into

a symbiotic relation with the plant: the bacteria obtain nourishment from the plant and in turn provide the plant with usable nitrogen in the form of ammonia (NH_3). The relation is an intimate one, in which the bacteria actually enter the roots of the legume and form the visible growths called nodules.

For the Romans a population of *Rhizobium* bacteria in the soil ensured root nodulation and subsequent nitrogen fixation in new fields of legumes. It is also

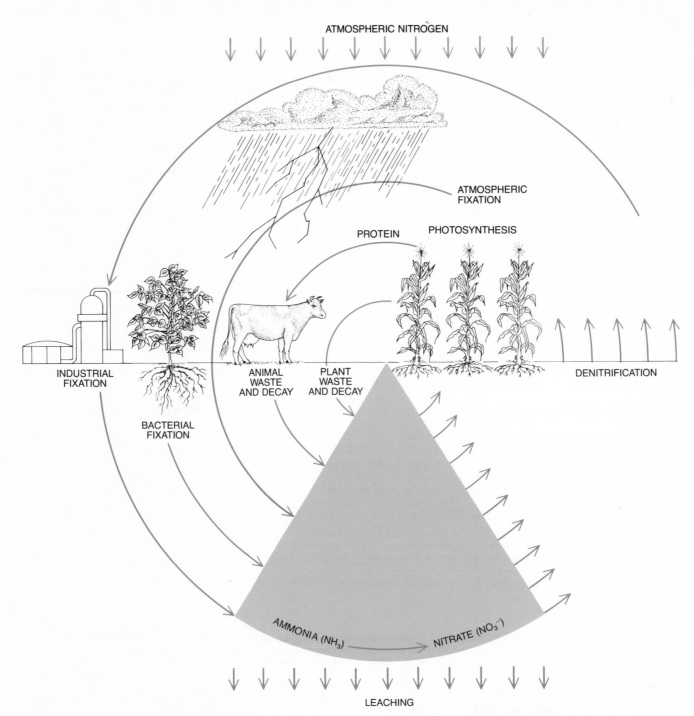

NITROGEN CYCLE maintains a balance between two vast reservoirs of nitrogen compounds: the earth's atmosphere and the earth's crust. Because green plants can utilize nitrogen only when it has been incorporated into chemical compounds such as ammonia (NH_3) the plants cannot extract nitrogen directly from the atmosphere, where it takes the form of diatomic molecules (N_2). Atmospheric nitrogen must therefore be fixed either industrially or by bacteriological or other natural processes such as lightning. Although only a small fraction of the total nitrogen supply is required by plants, fixation must be carried out constantly. Fixed nitrogen is lost through the leaching of the soil, through the harvesting of crops and through the action of denitrifying bacteria that convert fixed nitrogen into its diatomic form.

the basis of the traditional practice of crop rotation, because fixed nitrogen left in the soil by a crop of legumes can be taken up by grain plants, which do not form nodules. In 1888 *Rhizobium* was isolated by the German investigators Hermann Hellriegel and H. Wilfarth, and within 15 years deliberate inoculation of soil with cultured bacteria became agricultural practice. Today strains of *Rhizobium* are packaged with a supporting agent such as peat.

The importance of nitrogen to plant metabolism is now well understood biochemically. Nitrogen is incorporated into a great variety of biological molecules, and it is an essential constituent of proteins, where the peptide bond that links one amino acid to the next is formed between a nitrogen atom and a carbon atom. If the amino acids of dead plants were simply returned to the soil, new crops could recycle the dead matter immediately into new proteins. In the soil, however, amino acids are dismantled to yield ammonia or nitrate ions (NO_3^-). The nitrates are further broken down by the bacteria called denitrifiers into molecular nitrogen (N_2), which is returned to the atmosphere, completing a nitrogen cycle. Moreover, when crops are harvested or when rainwater carries dissolved nitrogen compounds to deeper levels of the ground, the nitrogen-bearing matter is physically removed from the topsoil. Leaching, harvesting and the activities of denitrifying bacteria result in a net loss of fixed nitrogen, which must be replaced if the next crop is to synthesize more protein for growth.

How can the nitrogen-fixing action of *Rhizobium* on legume roots be enhanced, thereby increasing crop yield? A straightforward approach is through plant breeding, which does not require microbiological techniques at all. For example, if the rate of photosynthesis were increased by selective breeding, the bacteria in the nodules might be able to fix more nitrogen. By combining plant breeding with microbiological techniques for modifying the *Rhizobium,* however, even higher yields of plant protein might be obtained.

A few years ago my colleagues and I at the University of Wisconsin at Madison began to apply the screening procedures of the pharmaceutical industry to nitrogen-fixing bacteria. The bacteria were first exposed to mutagenic substances or to ionizing radiation as a means of increasing the rate of mutation in the colony. Plants were then inoculated with mutant bacteria and the amount of nitrogen fixed by each strain of bacterium was measured. In this manner we obtained several mutants capable of significantly higher levels of nitrogen fixation than those afforded by standard inoculants. The soybean plants in the experiment showed superior growth.

The experiment was done in a laboratory growth chamber, and it was important to test the effects of the mutant bacteria on plant growth under field conditions. Soybeans were planted in nitrogen-poor Wisconsin fields; one plot in each field was inoculated with the mutant bacteria and a second was not inoculated. There was no difference in growth or yield between the two plots: even the uninoculated control plot produced a good soybean crop.

We had encountered precisely the effect that distinguishes agricultural from

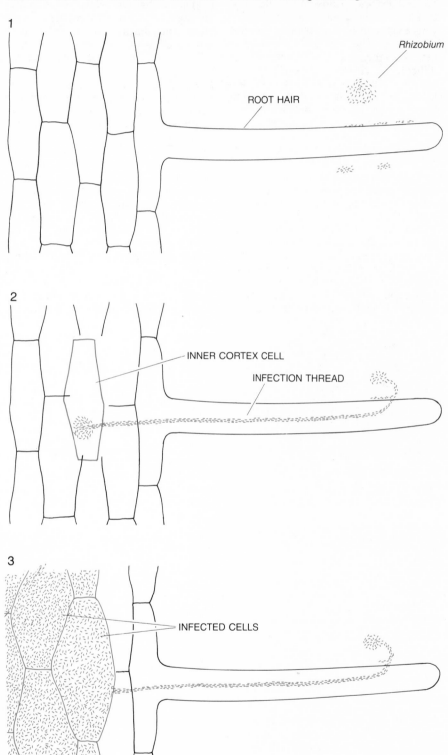

INFECTION OF A ROOT HAIR of a legume by *Rhizobium* begins when the bacteria attach themselves to the root hair (*1*) by means of a template-matching mechanism. The plant and the bacteria recognize each other through specific proteins. The bacteria then enter the root hair through an infection thread and stream into an internal cell of the root (*2*). The infection causes the cell to swell and divide (*3*). The result is a root nodule: a thick mass of infected cells.

industrial practice: the effect of natural populations of microorganisms in the uncontrolled environment of the open field. Indigenous strains of *Rhizobium* formed most of the root nodules in both test plots; the superior nitrogen-fixing mutants were unable to compete with the indigenous strains. When the mutant bacteria were introduced into fields where legumes had never been cultivated, they did lead to greater yields of soybeans. Our current plan to overcome the difficulty borrows a traditional tactic of the plant breeder: we seek to identify the most competitive strains of bacteria and to obtain mutants with better nitrogenfixing properties from those strains.

The procedure of inducing random mutations and then screening for useful mutant bacteria is inefficient. Microbiologists in a number of laboratories are studying the genetics and the biochemistry of infection by *Rhizobium* so that the bacteria can be directly modified by genetic engineering. In all nitrogen-fixing organisms the agent responsible for fixation is the enzyme nitrogenase, which catalyzes the conversion of molecular nitrogen into ammonia. In the reaction a transport protein donates electrons to the nitrogenase, which in turn transfers them to the diatomic molecule of nitrogen by a mechanism that is not yet fully understood. Three negatively charged electrons come to be associated with each atom of nitrogen; thereafter three protons (hydrogen nuclei) are withdrawn from the intracellular medium to neutralize the charge. Hence each diatomic molecule of nitrogen yields two molecules of ammonia.

The transfer of electrons from nitrogenase to molecular nitrogen has an energetically wasteful side reaction. Many of the electrons recombine with protons before they reach the nitrogen; the recombined electrons and protons are released as molecular hydrogen gas (H_2). Certain strains of *Rhizobium* synthesize hydrogenase, an enzyme that converts molecular hydrogen back into electrons and protons for reuse by nitrogenase. Hydrogenase could therefore serve as a kind of afterburner that would make the nitrogen-fixing bacteria more energyefficient. The enhanced bacterial efficiency would enable the plant to direct its energy more to seed yield than to the support of its symbiotic bacteria. Investigators at Oregon State University have recently demonstrated the effectiveness of hydrogenase in the open field. They showed that field-grown soybeans inoculated with a strain of *Rhizobium* that makes hydrogenase have a greater yield than soybeans inoculated with a strain lacking the enzyme.

Workers at the John Innes Institute in Britain have shown that the hydrogenase made by certain strains of *Rhizobium* is coded for by a single gene and that the gene is found on a plasmid, a loop of DNA separate from the bacterial chromosome. It should be possible to transfer the hydrogenase gene to *Rhizobium* strains that lack the enzyme but have other characteristics making them desirable as nitrogen fixers.

Not all symbiotic nitrogen fixation is conducted by *Rhizobium* bacteria, and not all nitrogen-fixing bacteria attach themselves to legumes. The key to engineering microorganisms that fix nitrogen for cereal crops may be in understanding the details of such natural symbioses. There is only one well-documented instance in which *Rhizobium* has been found to nodulate a plant other than a legume; it was reported by an investigator in Western Australia. A number of other nitrogen-fixing bacteria, however, have been identified and isolated. The bacterium *Frankia alni* fixes nitrogen for the alder tree and certain other nonlegume plants. Alder can therefore be employed in crop rotation much as legumes are: alder seedlings are mixed into forests in order to enrich the soil for commercially valuable trees such as Douglas fir and poplar.

Some bacteria fix nitrogen in the soil without entering into symbiosis with a plant. In my laboratory Stephen W. Ela and I are trying to get the nitrogen-fixing bacterium *Azotobacter vinelandii*, which does not take part in natural symbiotic relations, to bind to the roots of corn and so fertilize the corn. Ordinarily *A. vinelandii* makes no more ammonia than it needs for growth. Hence our first step was to obtain mutants that excrete excess ammonia. This can be accomplished by finding mutants for which feedback pathways that report an excess accumulation of ammonia have been blocked.

To make the ammonia available only to the corn our next goal has been to make the bacterium bind tightly to the roots of the corn plant. Several years ago two of my colleagues transferred certain *Rhizobium* genes to *A. vinelandii* and succeeded in getting the latter to stick tightly to clover roots. If genes that specify binding to corn roots rather than clover roots can be introduced into *A. vinelandii*, the corn should be able to take in the ammonia excreted by the bacteria.

In another aspect of the same project we are attempting to breed varieties of corn that will be able to meet the ener-

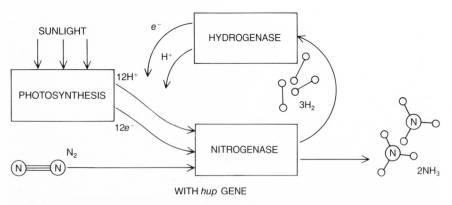

BIOCHEMICAL SHORT CIRCUIT leads to a wasteful release of hydrogen gas (H_2) as a byproduct of the enzymatic reaction by means of which *Rhizobium* bacteria convert molecular nitrogen (N_2) into ammonia (NH_3). The hydrogen has no known value to the plant or to the bacteria; instead it seems merely to waste energy derived from photosynthesis. The energy serves to segregate free protons (H^+) and electrons (e^-), which then drive the reactions of nitrogen fixation. If the protons and electrons recombine to form hydrogen, however, the energy is dissipated. In some strains of *Rhizobium* bacteria a gene carried by a plasmid circumvents this inefficiency. The gene, designated *hup*, codes for the synthesis by the bacteria of the enzyme hydrogenase. Hydrogenase catalyzes the breakdown of hydrogen gas into its constituent protons and electrons for reapplication in nitrogen fixation. The introduction of the *hup* gene into a *Rhizobium* inoculant has been shown to give the inoculated plant higher seed yields than those of plants infected with strains of *Rhizobium* bacteria that do not synthesize hydrogenase.

gy requirements of the bacterium. The corn plants currently under cultivation in the U.S. cannot support *A. vinelandii*. By selective breeding of varieties of corn from throughout the world Ela has been able to increase the production on corn roots of carbon compounds that serve as a source of energy and electrons for nitrogen fixation by *A. vinelandii*. We have now developed corn plants capable of obtaining perhaps 1 percent of their nitrogen from the bacterial association, and we are sufficiently encouraged by our results to try to improve the percentage.

Plants can benefit from many other associations with microorganisms that are just beginning to be understood. At the University of California at Berkeley biologists have shown that the addition of certain strains of the bacterium *Pseudomonas putida* to sugar beet seeds or to potatoes increases the yield of the plant. It appears the bacterium secretes agents that sequester iron in the soil. By this means the iron near the roots of the plant is made unavailable to potentially harmful fungi and bacteria that need iron for growth.

The soil fungi called mycorrhizae colonize plant roots and can create what amounts to an extension of the root system of the plant. For example, they can make phosphate available to plants in phosphate-poor soil by converting the phosphate into a soluble form and transporting it to the roots of the plants. Mycorrhizae can also transport water to the plant, collected beyond the reach of the plant's root system. Plants grown on land reclaimed from strip-mining have been inoculated with strains of mycorrhizae. Other strains may soon have considerable economic importance in forestry because they stimulate the growth of tree seedlings. So far, however, there has been relatively little work done to match specific strains of mycorrhizae to specific plants and growing conditions.

The culture of individual plant cells or groups of cells can provide a source of plant products that is not subject to variability of crop yields or to the uncertainties of international trade. Many plant products, such as pharmaceuticals, pesticides and flavoring agents, have been isolated from tropical plants and can probably be produced by growing plant cells in large fermenters. Moreover, the individual plant cell is an extraordinarily efficient and convenient medium in which to develop new plant varieties. When plant cells are exposed to a mutagen and placed in a stressful environment, varieties adapted to the stress quickly appear and can be selected for culturing.

When a stress such as a toxic agent or the lack of an essential nutrient is imposed, only those mutant cells that by chance are adapted to the stress will survive. In successive generations of cells the progeny of the selected cells can be developed into finely tuned organisms adapted to a specific set of environmental conditions. The same method has long been employed in selecting industrial microorganisms such as molds and bacteria that can synthesize antibiotic agents resistant to degradation.

Once the desired plant cell has been selected it can be grown in culture to form the mass of unorganized tissue called a callus. Plant hormones can sometimes cause the callus to become organized into the stems, roots and other differentiated parts of the mature plant. A plant resistant to the fungal toxin that causes southern corn leaf blight was developed from tissue culture by biologists at the University of Minnesota. They applied the toxin to the culture and then selected the resistant cell lines.

Only a few plants have been successfully regenerated from single cells, and the regenerated plants do not always inherit the properties selected for in the single cell. A herbicide-resistant cell does not necessarily give rise to a herbicide-resistant plant, even though the individual cells remain herbicide-resistant when they are grown again in culture. Moreover, single plant cells grown in culture are usually diploid or polyploid: each cell has at least two copies of every chromosome. The genetic information carried by the cell is therefore coded at least twice. In these circumstances most mutations are recessive. The recessive gene has no effect on the parent plant, but it may make its presence felt in succeeding generations. As a result the characteristics of the progeny of the cells cannot be predicted with certainty. Haploid cells, which have only one copy of each chromosome, have now begun to be cultured, and it should be easier to detect the mutants.

AUGMENTED ROOT SYSTEM is formed when loblolly pine seedlings are inoculated with the soil fungus *Pisolithus tinctorius*, a species of mycorrhiza. In the upper photograph the fungi are absent and the surface area of the roots is limited. In the lower photograph the mycorrhizae contribute to a denser root system with a larger surface area for the absorption of water and nutrients. The inoculated seedlings grow more rapidly and are more likely to survive. Colonies of mycorrhizae can also extend into regions of the soil not accessible to the root system of a plant, thereby increasing the volume of the soil that can be tapped by the root system.

Removing the walls of a plant cell with enzymes gives rise to the naked cell called a protoplast. Protoplasts from two unrelated plants can be made to fuse, creating a single cell that can regenerate a cell wall and grow for many generations in a nutrient solution. In a sense the process is equivalent to sexual reproduction between different species of plants, but a mature hybrid plant is seldom generated. At the Max Planck Institute for Biology in Tübingen protoplasts from a potato and a tomato were fused and did develop into a mature hybrid plant that has been given the name pomato. The pomato plant is sterile, however, and bears neither potatoes in the ground nor tomato fruits hanging from the stems.

The transfer of genes from one organism to another represents the most sophisticated application of microbiological strategies to agriculture, but it is the least well developed of the three I have mentioned. Indeed, compared with the recombinant-DNA methods developed for work with animal genes the work done with plant genes is quite limited. The principles of foreign gene insertion, however, are the same for plant cells as they are for animal cells and for bacterial cells.

In order to insert a plant gene into a bacterium it is first necessary to isolate the gene by means of the enzymes known as restriction endonucleases. The target gene must then be transported inside the host cell by a plasmid or virus vector. In this way workers in a number of laboratories have cloned plant genes in the bacterium Escherichia coli. The successful introduction of a foreign gene into a cell, however, does not always mean that the protein product of the gene will be expressed by the cell.

For proteins to be expressed additional intracellular chemistry must be set in motion, and it is not yet understood precisely how this can be accomplished in every case.

If microbiologists succeed in expressing plant proteins in bacteria, the proteins can be made by growing the bacteria in fermentation tanks. The process could then be carried one step further, at least in principle. Numerous valuable plant products such as pharmaceuticals, pesticides, oils, waxes and flavoring agents are generally synthesized by plants in multistage chemical reactions where several enzymes take part. If the genes for all the enzymes can be expressed in a bacterium, the bacterium can in effect become a factory for the synthesis of the plant compound.

A much more challenging goal is the introduction of foreign genes into plant cells. Several potential paths are being explored. Perhaps the most promising method focuses on a plasmid found in the bacterium Agrobacterium tumefaciens. A. tumefaciens induces a tumor called a crown gall in wounded dicotyledons, the large class of flowering plants that includes legumes, tomatoes and numerous other crop plants (but not the cereal grains). The mechanism of the infecting bacterium is to insert a segment of its plasmid into a chromosome of the plant cells. The inserted segment is called transfer DNA, or T-DNA.

The insertion of T-DNA is therefore a natural form of genetic modification, and it endows the infected plant cells with several unusual properties that are probably essential to the formation of a crown gall. Ordinary cells proliferate in culture only in the presence of plant growth hormones, but cells infected with A. tumefaciens do not require such

hormones. The release of the cells from hormonal control may account for their unusually rapid growth in the tumor. The infected cells also manufacture the enzyme opine synthetase, which catalyzes the synthesis by the plant cell of nitrogen-rich compounds called opines. Opines seem to be required by A. tumefaciens as a source of nitrogen. Hence crown galls can be understood as the outcome of a biological strategy developed by the bacterium to secure the nitrogen necessary for its growth.

Investigators at the University of Leiden have been able to infect tobacco cells in culture with A. tumefaciens. They showed that tobacco plants regenerated from the infected cells retained the T-DNA and continued to make opine synthetase. More recently investigators at the Max Planck Institute for Plant Breeding in Cologne demonstrated that the gene coding for the expression of opine synthetase is passed on through the seed to succeeding generations. Such results justify some confidence that if foreign genes can be spliced into the A. tumefaciens plasmid in association with the T-DNA, they will be expressed as proteins in the mature plant and carried through the seed to the progeny.

Another method under investigation for introducing foreign DNA into plant cells utilizes the cauliflower mosaic virus (CaMV). DNA from this plant virus can be isolated and spliced into a plasmid for insertion into the bacterium E. coli. There the DNA can be amplified, or reproduced many times; the amplified DNA retains the ability to infect the cauliflower and a range of plants related to it. Investigators are now working to determine which places along the viral DNA are suitable for the introduction of foreign genes.

FOREIGN DNA — ISOLATE

FOREIGN GENE

T-DNA

RECOMBINANT PLASMID — INFECT

PLANT CHROMOSOMES — REPLICATE

T-DNA — PLASMID — ISOLATE — CLEAVE

Agrobacterium tumefaciens

INTRODUCTION OF FOREIGN DNA into plant cells may be accomplished by exploiting the natural infection process of the bacterium *Agrobacterium tumefaciens*. The bacterium carries a plasmid (a loop of DNA separate from the bacterial chromosome) that causes crown-gall tumors in most dicotyledonous plants and induces the infected plant cells to synthesize the nitrogen compounds called opines. The mechanism of the infection has been termed genetic colonization: a section of the plasmid called T-DNA combines with chromosomal DNA in the nucleus of the plant cell. The plasmid might therefore serve as a vector for inserting foreign DNA into plant cells. The

If genes can be introduced into plant cells at will, one important application may be the insertion of nitrogen-fixing genes (*nif* genes) into cereal plants. Workers at the University of Sussex have assembled a bacterial plasmid that includes all 17 of the known *nif* genes from the nitrogen-fixing bacterium *Klebsiella pneumoniae*. When the plasmid was transferred to *E. coli,* the latter, which is normally incapable of fixing nitrogen, became a nitrogen-fixing microorganism.

Even more promising are the recent successes of groups of investigators at Cornell University, the Pasteur Institute and the University of Paris. They introduced the 17 *nif* genes of *K. pneumoniae* into yeast. Yeasts, being eukaryotic organisms whereas bacteria are prokaryotic, are much more closely related to the higher plants than they are to bacteria. Hence the introduction of *nif* genes into yeast cells marks the crossing of a significant biological barrier.

Nevertheless, the yeast cells carrying the *nif* genes were not able to express the inserted DNA: they were not able to fix nitrogen from the atmosphere. The failure illustrates the complexity associated with the genetic engineering of biological functions embodied in more than one gene. The transferred DNA must first be transcribed correctly into RNA by the yeast. Correct transcription cannot be assumed as a matter of course, because the yeast must correctly interpret the bacterial signals to start the transcription and to stop it. The RNA must then be exported from the nucleus and be recognized by the ribosomes as a messenger RNA for translation into protein. The 17 proteins that express the *nif* genes must then function together in the foreign cytoplasm of the yeast cell.

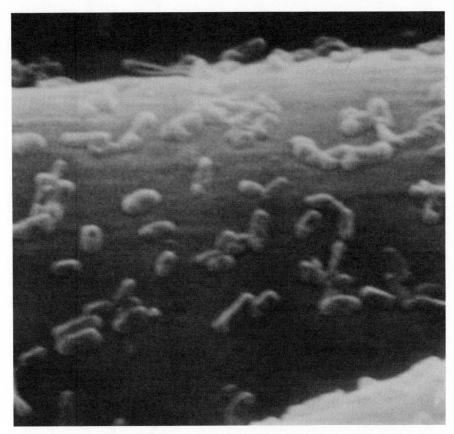

BENEFICIAL BACTERIA attached to the root hairs of a sugar beet plant are enlarged 6,300 diameters in a scanning electron micrograph made by Trevor V. Suslow and Douglas G. Garrott of the University of California at Berkeley. The bacteria *Pseudomonas putida* suppress the growth of other microorganisms near the roots by extracting iron from the soil. The iron is tightly bound in molecules synthesized by the symbiotic bacteria and so becomes unavailable to fungi and other soil bacteria that could be harmful to the plant roots. Competition for essential nutrients is thereby reduced and significantly higher sugar beet yields can result.

There may be impediments to such functioning. For example, the nitrogenase molecule has in its structure a large number of iron atoms. Enough iron is evidently on hand in nitrogen-fixing bacteria, but it is not certain the heavy demand for iron can be met in a plant cell without jeopardizing the synthesis of other enzymes essential to the plant.

Doubts of this kind can arise even

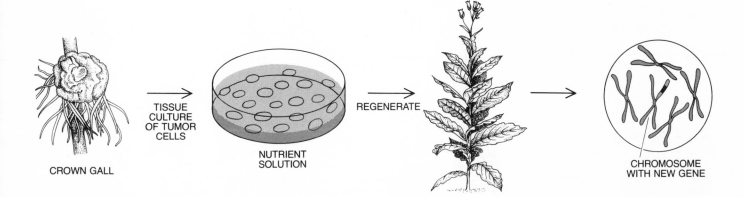

CROWN GALL → TISSUE CULTURE OF TUMOR CELLS → NUTRIENT SOLUTION → REGENERATE → CHROMOSOME WITH NEW GENE

plasmid would be cut open at a site within the T-DNA and the foreign gene would be spliced into it. The T-DNA is replicated when the tumor cells of a crown gall divide, and tumor cells grown in tissue culture continue to carry T-DNA. It has been possible in some cases to regenerate a plant from the cultured tumor cells; T-DNA is still found in the chromosomes of the regenerated plant. Moreover, the gene carried by T-DNA that codes for the enzyme opine synthetase is passed on to daughter plants as if it were an ordinary dominant gene. If foreign genes inserted into T-DNA are also transmitted to plant progeny, new plant strains could be genetically engineered.

when it seems at first that the expression of a particular plant function calls for only one gene. The enhancement of protein quality in food is a major goal of agricultural research. The stored proteins in some grains and other seeds, for instance, are deficient in amino acids essential to human and animal nutrition. It was once thought that the insertion of a single gene, coding for a stored protein with a better balance of amino acids, would be sufficient to improve the quality of the protein. It is now understood that major seed proteins are often collections of numerous related proteins, each of the latter being encoded by a separate gene. In order to make a significant improvement in the overall quality of the stored protein many of the genes might have to be individually modified.

Although such technical problems may be difficult to overcome, they are probably not the limiting factors in the application of microbiology to agriculture. In the long run more stringent constraints will be the comparatively small financial commitment to basic research in agriculture and the loss of irreplaceable genetic resources from the total gene pool.

According to an analysis prepared by the National Science Foundation, Federal support for research into all aspects of plant science was $206 million in the fiscal year 1977, 2.3 percent of the $8.8 billion spent by the Federal Government during that year for all basic and applied research. President Reagan's 1982 budget calls for a $691-million research allocation to the Department of Agriculture, a 5.3 percent share of the $13.1 billion now proposed for all basic and applied research in 1982. Of this, however, only a small fraction will be available for studies of agricultural applications of microbiology. For example, the Competitive Research Grants program of the Department of Agriculture will administer $26 million in basic research funds in 1982 for studies of nitrogen fixation, photosynthesis, genetic mechanisms of crop improvement, plant and environmental stress and the requirements of human nutrition. Microbiological studies will receive only $4.8 million. Of all recombinant-DNA investigations monitored by the National Institutes of Health, Federal grants for genetic engineering in agriculture amounted to only $1 million in the fiscal year 1980, compared with some $24.5 million for medical research and $27.5 million for general research.

The accelerated destruction of the gene pool is doubly ironic. It is caused primarily by the clearing of land in the tropical rain forest for farming. Moreover, it is happening at the dawn of an age in which such genetic wealth, until now a relatively inaccessible trust fund, is becoming a currency with high immediate value. Estimates place the annual extinction rate as high as 1,000 species per year, and there is still little organized effort for maintaining genetic reserves.

The microbiologist will be only one worker among many who will participate in the development of new field crops and agricultural practices. It is particularly important that the microbiologist work closely with the plant breeder, not only because of the experience of breeders with the effects of laboratory modifications on field crops but also because many of the techniques of the two disciplines actually overlap. Future work may disclose unexpected barriers to the application of microbiological ideas in agriculture, but it is at least as likely that such ideas can readily be put into practice.

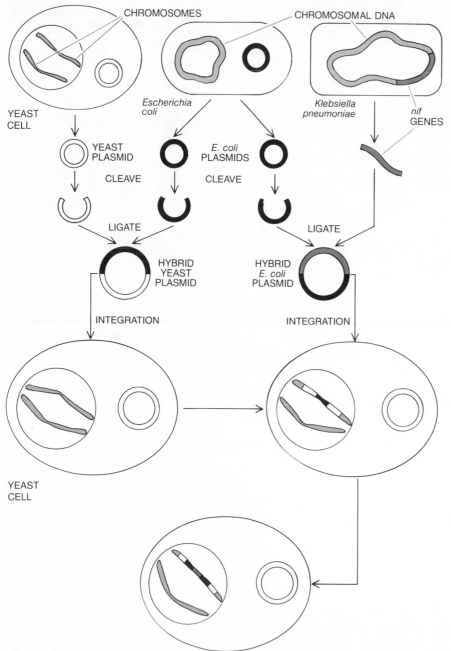

GENES FOR NITROGEN FIXATION have been inserted into the genome of a yeast in a two-stage process. In the first stage plasmids from the bacterium *Escherichia coli* and from a yeast cell are cleaved and then fused to form a single hybrid plasmid. The hybrid plasmid can be recognized by the yeast cell and integrated into its chromosomal DNA. In the second stage the genes to be introduced into the yeast are isolated from the chromosome of the bacterium *Klebsiella pneumoniae*, a nitrogen-fixing organism. The genes, collectively designated *nif*, code for some 17 proteins. Another *E. coli* plasmid is cleaved and the isolated *nif* genes are introduced to form a second hybrid plasmid. Because of the bacterial DNA already inserted into one of the yeast chromosomes the yeast cell recognizes the hybrid *E. coli* plasmid. The plasmid is then integrated into the yeast chromosome. The experiment was carried out by Aladar A. Szalay and his co-workers at Cornell University. Although the insertion of the prokaryotic *nif* genes into the eukaryotic yeast cells demonstrates that genetic material can be transferred between different biological systems, the nitrogen-fixing proteins are not expressed in yeast.

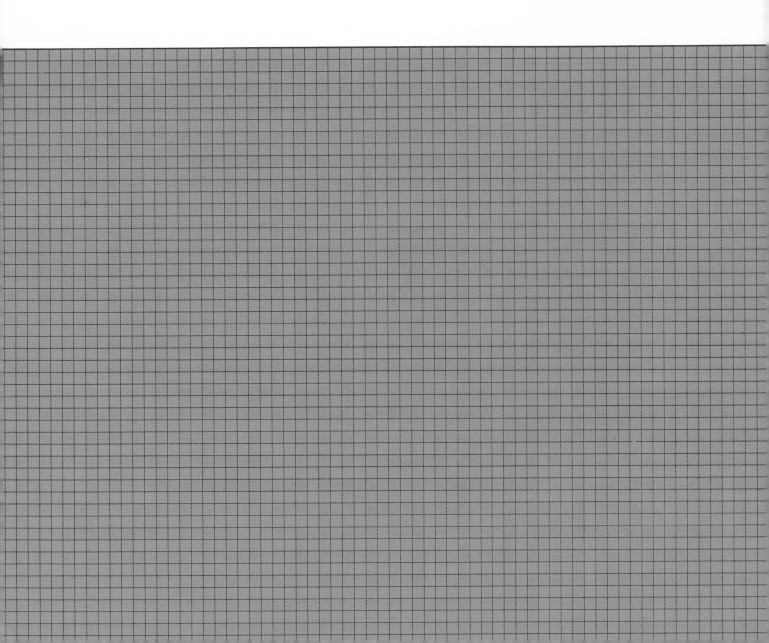

The Authors

ARNOLD L. DEMAIN and NADINE A. SOLOMON ("Industrial Microbiology") are microbiologists at the Massachusetts Institute of Technology. Demain, who is professor of industrial microbiology in the department of nutrition and food science, majored in bacteriology as an undergraduate and graduate student at Michigan State College (now Michigan State University). He then went to the University of California, where he studied at both the Davis and the Berkeley campuses, obtaining his Ph.D. in microbiology in 1954. Before joining the M.I.T. faculty in 1969 he was head of the department of fermentation research at the Merck, Sharp & Dohme Research Laboratories. An active member of the American Society for Microbiology, Demain has served on numerous committees having to do with applied microbiology and fermentation technology. Solomon, who is a research associate in the department of nutrition and food science at M.I.T., is a graduate of the University of Massachusetts. She has worked with Demain since 1972 and in recent years has collaborated with him on a number of journal articles in their field.

HERMAN J. PHAFF ("Industrial Microorganisms") has a joint appointment as professor of food science and technology and professor of bacteriology at the University of California at Davis. A native of the Netherlands, he was trained as a chemical engineer at the Technical University of Delft. He came to the U.S. in 1939 to continue his studies, receiving a Ph.D. from the University of California at Berkeley in 1943. He taught at Berkeley until 1951, when he moved to the Davis campus. A specialist in the ecology and molecular taxonomy of yeasts, he is the coauthor (with Martin W. Miller and Emil M. Mrak) of *The Life of Yeasts*. In addition to Phaff's professional activities he is an accomplished cellist and participates in many musical events on the Davis campus.

DAVID A. HOPWOOD ("The Genetic Programming of Industrial Microorganisms") is John Innes Professor of Genetics at the University of East Anglia. He is also head of the genetics department at the nearby John Innes Institute, a research establishment supported by the British Agricultural Research Council. Hopwood studied natural sciences, specializing in botany, at the University of Cambridge. Soon after his graduation in 1954, he writes, "I began research for my doctorate at Cambridge, studying the genetics of *Streptomyces,* about which little was known at the time. I am still working on the same topic; it gets more fascinating all the time." He obtained a Ph.D. from Cambridge in 1958 and a D.Sc. from the University of Glasgow in 1974. He was a lecturer in genetics at Glasgow from 1961 to 1968, when he took up his present positions. In 1979 Hopwood was elected a Fellow of the Royal Society.

ANTHONY H. ROSE ("The Microbiological Production of Food and Drink") is professor of microbiology at the University of Bath. His degrees, in applied biochemistry, are from the University of Birmingham. After postdoctoral work at Rutgers University and the National Research Council of Canada in Ottawa he taught at Heriot-Watt University in Edinburgh and at the University of Newcastle upon Tyne. He joined the faculty at Bath in 1968. His main research interest for the past decade or so has been in the composition and function of the envelope of *Saccharomyces cerevisiae* (brewer's yeast). This is Rose's fourth article for SCIENTIFIC AMERICAN; the previous ones were "Beer" (June, 1959); "Yeasts" (February, 1960) and "New Penicillins" (March, 1961).

YAIR AHARONOWITZ and GERALD COHEN ("The Microbiological Production of Pharmaceuticals") are microbiologists on the faculty of Tel Aviv University. Aharonowitz got his Ph.D. from Tel Aviv in 1974. He then spent two years as a research associate at the Massachusetts Institute of Technology, where, he writes, "I worked with Arnold L. Demain, who introduced me to the field of industrial microbiology. In Demain's laboratory I studied different aspects of the metabolic regulation of beta-lactam antibiotic production in streptomycetes." Aharonowitz joined the department of microbiology at Tel Aviv in 1976. In addition to his work on biochemical and genetic regulatory mechanisms, he reports, "I am involved in several aspects of the industrial application of biocatalysis." Cohen is a graduate of University College London; after obtaining his B.Sc. degree he took a year off to work on a kibbutz in Israel. He later joined the polymer department of the Weizmann Institute of Science in Rehovot, where he was awarded his doctorate in 1968. He spent the next three years in the U.S. working at the National Institutes of Health before he returned to Israel in 1971 to join the department of microbiology at Tel Aviv. In recent years, he writes, "I have become involved in the area of applied microbiology, and in particular in the ways microorganisms can be exploited to make useful amounts of vitamins, amino acids and antibiotics."

DOUGLAS E. EVELEIGH ("The Microbiological Production of Industrial Chemicals") is professor of microbiology at Cook College/New Jersey Agricultural Experiment Station, a division of Rutgers University. Born and educated in Britain, he received his Ph.D. from the University of Exeter in 1959. Before joining the Rutgers faculty in 1970 he worked for six years at the Canadian National Research Council's Prairie Regional Laboratory in Saskatoon. In his spare time, Eveleigh writes, he "rejuvenates on table tennis and legerdemain."

ELMER L. GADEN, JR. ("Production Methods in Industrial Microbiology"), is Wills Johnson Professor of Chemical Engineering at the University of Virginia. His degrees are from Columbia University: B.S., 1944; M.S., 1947; Ph.D., 1949. He joined the faculty at Columbia soon after obtaining his doctorate and taught there until 1974, when he left to become dean of the College of Engineering, Mathematics and Business Administration at the University of Vermont. He was appointed to his present position in 1979. Gaden's primary professional interest is in the development of biochemical processes for the production of food and chemicals and for the disposal of wastes. In 1970 he was the first recipient of the Food and Bioengineering Award of the American Institute of Chemical Engineers. In recent years, Gaden reports, he has been particularly active in the work of the Board of Science and Technology for International Development of the National Academy of Sciences/National Research Council.

WINSTON J. BRILL ("Agricultural Microbiology") is Vilas Research Professor of Bacteriology at the University of Wisconsin at Madison. A graduate of Rutgers University, he got his Ph.D. in microbiology from the University of Illinois at Urbana-Champaign in 1965. After two years of postdoctoral work on the genetics and regulation of amino acid metabolism at the Massachusetts Institute of Technology he moved to Wisconsin, where his research is devoted primarily to the study of the biochemistry, genetics and physiology of nitrogen fixation [see "Biological Nitrogen Fixation," by Winston J. Brill; SCIENTIFIC AMERICAN, March, 1977]. In addition to his academic work he has just been made director of research for the Cetus Madison Corporation, a newly formed subsidiary of the Cetus Corporation that will concentrate on the application of modern microbiological techniques to agriculture.

Bibliographies

Readers interested in further explanation of the subjects covered by the articles in this issue may find the following lists of publications helpful.

INDUSTRIAL MICROBIOLOGY

ECONOMIC MICROBIOLOGY: VOL. 1, ALCOHOLIC BEVERAGES; VOL. 2, PRIMARY PRODUCTS OF METABOLISM; VOL. 3, SECONDARY PRODUCTS OF METABOLISM; VOL. 4, MICROBIAL BIOMASS; VOL. 5, MICROBIAL ENZYMES AND BIOCONVERSIONS. Edited by Anthony H. Rose. Academic Press, 1977–80.

INDUSTRIAL MICROORGANISMS

BIOLOGY OF MICROORGANISMS. Thomas D. Brock. Prentice-Hall, Inc., 1970.

THE MICROBIAL WORLD. Roger Y. Stanier, Edward A. Adelberg and John L. Ingraham. Prentice-Hall, Inc., 1976.

THE LIFE OF YEASTS. H. J. Phaff, M. W. Miller and E. M. Mrak. Harvard University Press, 1978.

INTRODUCTORY MYCOLOGY. Constantine J. Alexopoulos and Charles W. Mims. John Wiley & Sons, Inc., 1979.

FUNDAMENTALS OF HUMAN LYMPHOID CELL CULTURE. J. Leslie Glick. Marcel Dekker, Inc., 1980.

MICROBIOLOGY OF FOODS. John C. Ayres, J. Orvin Mundt and William E. Sandine. W. H. Freeman and Company, 1980.

THE GENETIC PROGRAMMING OF INDUSTRIAL MICROORGANISMS

THE MOLECULAR BASIS OF MUTATION. John W. Drake. Holden-Day, Inc., 1970.

MOLECULAR BIOLOGY OF THE GENE. James D. Watson. W. A. Benjamin, Inc., 1976.

THE MANY FACES OF RECOMBINATION. D. A. Hopwood in *Proceedings of the Third International Symposium on Genetics of Industrial Microorganisms,* edited by O. K. Sebek and A. I. Laskin. American Society for Microbiology, 1979.

PLASMIDS. P. Broda. W. H. Freeman and Company, 1979.

GENETIC ENGINEERING: PRINCIPLES AND METHODS. Edited by J. K. Setlow and Alexander Hollaender. Plenum Press, 1980.

FRESH APPROACHES TO ANTIBIOTIC PRODUCTION. D. A. Hopwood and K. F. Chater in *Philosophical Transactions of the Royal Society of London, Series B,* Vol. 210, No. 1040, pages 313–328; August 11, 1980.

THE MICROBIOLOGICAL PRODUCTION OF FOOD AND DRINK

MICROBIAL PRODUCTS IN FOODS: SYMPOSIUM CONVENED BY K. S. KANG, in *Developments in Industrial Microbiology,* Vol. 19, pages 69–131; 1978.

MICROBIAL TECHNOLOGY: VOL. 1, MICROBIAL PROCESSES; VOL. 2, FERMENTATION TECHNOLOGY. Edited by H. J. Peppler and D. Perlman. Academic Press, 1979.

MICROBIOLOGY OF FOOD FERMENTATIONS. C. S. Pederson. Avi Publishing Co., 1979.

THE MICROBIOLOGICAL PRODUCTION OF PHARMACEUTICALS

APPLICATIONS OF BIOCHEMICAL SYSTEMS IN ORGANIC CHEMISTRY. Edited by J. Bryan Jones, Charles J. Sih and D. Perlman. John Wiley & Sons, Inc., 1976.

ANTIBIOTICS AND OTHER SECONDARY METABOLITES: BIOSYNTHESIS AND PRODUCTION. Edited by R. Hutter, T. Leisinger, J. Nüesch and W. Wehrli. Academic Press, 1978.

PRINCIPLES OF GENE MANIPULATION. R. W. Old and S. B. Primrose in *Studies in Microbiology: Vol. 2.* University of California Press, 1980.

CONTROL OF ANTIBIOTIC BIOSYNTHESIS. J. F. Martin and A. L. Demain in *Microbiological Reviews,* Vol. 44, No. 2, pages 230–251; June, 1980.

RECOMBINANT DNA. Special issue of *Science,* Vol. 209, No. 4463; September 19, 1980.

THE MICROBIOLOGICAL PRODUCTION OF INDUSTRIAL CHEMICALS

PLASMIDS OF MEDICAL, ENVIRONMENTAL AND COMMERCIAL IMPORTANCE. Edited by K. N. Timmis and A. Pühler. Elsevier North-Holland, Inc., 1979.

FERMENTATION: SCIENCE AND TECHNOLOGY WITH A FUTURE. Edited by A. I. Laskin, M. C. Flickingers and E. L. Gaden, Jr., in *Biotechnology and Bioengineering,* Vol. 22, Supplement; 1980.

IMPACTS OF APPLIED GENETICS: MICROORGANISMS, PLANTS AND ANIMALS. Office of Technology Assessment. U.S. Government Printing Office, 1981.

TRENDS IN THE BIOLOGY OF FERMENTATION FOR FUELS AND CHEMICALS. Edited by A. Hollaender, R. Rabson, P. Rogers, A. San Pietro, R. Valentine and R. Wolfe. Plenum Press, 1981.

PRODUCTION METHODS IN INDUSTRIAL MICROBIOLOGY

PRINCIPLES OF MICROBE AND CELL CULTIVATION. S. John Pirt. John Wiley & Sons, Inc., 1975.

INDUSTRIAL MICROBIOLOGY. Brinton M. Miller and Warren Litsky. McGraw-Hill Book Company, 1976.

FERMENTATION AND ENZYME TECHNOLOGY. Edited by Daniel I.-C. Wang, C. L. Cooney, A. L. Demain, P. Dunnill, A. E. Humphrey and M. D. Lilly. John Wiley & Sons, Inc., 1979.

AGRICULTURAL MICROBIOLOGY

NITROGEN FIXATION: BASIC TO APPLIED. Winston J. Brill in *American Scientist,* Vol. 67, No. 4, pages 458–466; July–August, 1979.

GENETIC IMPROVEMENT OF CROPS. Edited by Irwin Rubenstein, Burle Gengenbach, Ronald L. Phillips and C. Edward Green. University of Minnesota Press, 1980.

GENOME ORGANIZATION AND EXPRESSION IN PLANTS, 1979. NATO Advanced Study Institute on Genome Organization and Expression in Plants. Edited by C. J. Leaver. Plenum Press, 1980.

PERSPECTIVES IN PLANT CELL AND TISSUE CULTURE. Edited by Indra K. Vasil. Academic Press, 1980.

Index